BASIC SEAMANSHIP

Also by PETER CLISSOLD

Radar in Small Craft
Yacht Master: Offshore and Ocean
Brown's Star Chart (with A. J. Tweddell)
Marista Terminaro
The "Shield" Protractor

BASIC
SEAMANSHIP

BY

PETER CLISSOLD

Commander R.N.R. (Retd.) Master Mariner
Younger Brother of Trinity House
Fellow of the Royal Institute of Navigation

REVISED BY

Captain A. G. W. MILLER, Extra Master

GLASGOW
BROWN, SON & FERGUSON, LTD., Publishers
4-10 DARNLEY STREET

First Edition   1936
Sixth Edition   1975
Seventh Edition   1998

ISBN 0 85174 650 0
ISBN 0 85174 255 6 (Sixth Edition)

© 1998 Brown, Son & Ferguson, Ltd., Glasgow, G41 2SD
Printed and made in Great Britain

Dedicated
To all those who are sent to fetch
The Key of the Keelson.

FOREWORD

This book is intended for anyone interested in boats and ships and to show something of the seaman's art, so far as this can be reduced to writing.

This edition, the seventh, has again been enlarged. Though not a text book, it has been designed to cover the requirements for the following examinations.

D.Tp.   Able Seaman's Certificate,
                        Lifeboatman's Certificate,
R.Y.A.   National Coastal Certificate.
R.Y.A./D.Tp.   Yachtmaster's Certificate (Offshore and Ocean).

Permission has been received to produce the Royal Naval code flags and extracts of recommended safety equipment for small pleasure craft from the Controller of H.M. Stationery Office.

For permission to reproduce photographs I am grateful to:

Schat-Davit Company (Lifeboats & Davits)

Elliot Equipment Ltd. (Inflatable Liferaft)

British Ropes Ltd. (Splicing Braided Rope)

Photographic Section H.M.S. *Invincible* (Replenishing at Sea)

P&O European Ferries (*Pride of Bilbao*)

London Sailing Project (*Rona II*, Youth Training Ship)

The Royal Yacht Squadron; the Royal Thames Yacht Club; the Royal Clyde Yacht Club; the Royal Cruising Club; the Royal Ocean Racing Club (Yacht Club Burgees).

The best of the illustrations have been drawn by Commander C. H. Williams, R.D., R.N.R.

A

"A's for the Anchor we oft times let go".

Aback — The position of the sails when the wind is pressing on their fore side.

Abaft — Behind; nearer the stern than, *e.g.*, abaft the beam, abaft the foremast.

Abeam — Abreast on the beam; at right angles to the ship's fore-and-aft line.

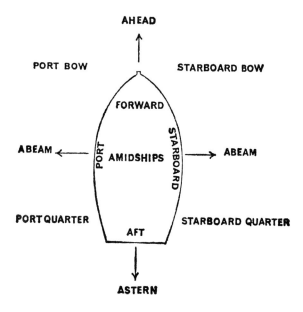

Able-Seaman — From "Able-bodied Seaman"; shortened to A.B. A skilled seaman. A man must pass an examination before being rated A.B. in the Merchant Navy, candidates must be at least 18 years old; have 12 months' service in a deck rating at sea (subject to a remission for pre-sea training) or 18 months in a general purpose (G.P.) rating, have a certificate of Proficiency in Survival Craft; have a Steering Certificate.

The examination will be held on the following Syllabus:

NAUTICAL KNOWLEDGE

(1) The meaning of common nautical terms.

(2) The names and functions of various parts of the ship (*e.g.*, decks, compartments, ballast tanks, bilges, air pipes, strum boxes, etc.).

(3) Knowledge of the compass card 0° to 360°. Ability to report the approximate bearing of an object in degrees or points on the bow.

(4) Understanding helm orders.

(5) Reading, streaming and hauling (handing) in a patent log.

(6) Markings on a hand lead line, taking a cast of the hand lead and correctly reporting the sounding obtained.

(7) Marking of the anchor cable.

(8) The use of life saving and fire-fighting appliances.

PRACTICAL WORK

Tested as far as possible by practical demonstration

(9) Knots, hitches and bends in common use:

Reef Knot	Bowline and bowline on the bight.
Timber hitch	Sheet bend, double and single.
Clovehitch	Sheepshank.
Rolling hitch	Round turn and two half hitches.
Figure of eight	Marlinespike hitch.
Wall and crown.	

To whip a rope's end using plain or palm and needle whipping. To put a seizing on rope and wire. To put a stopper on a rope or wire hawser, and derrick lift.

(10) Splicing plaited and multi-strand manila and synthetic rope, eye splice, short splice and back splice. Splicing wire rope, eye splicing using a locking tuck. Care in use of rope and wire.

(11) Slinging a stage, rigging a bosun's chair and pilot ladder.

(12) Rigging a derrick. Driving a winch, general precautions to be taken before and during the operation of a winch whether used for working cargo or for warping.

(13) The use and operation of a windlass in anchor work and in warping. Safe handling of moorings with particular reference to synthetic-fibre ropes and self tensioning winches. Precautions to be taken in the stowage of chain cable and securing the anchors for sea.

(14) A knowledge of the gear used in cargo work and an understanding of its uses. General maintenance with particular reference to wires, blocks and shackles.

(15) The safe handling of hatch covers including mechanical hatch covers, battening down and securing hatches and tank lids.

(16) If no lifeboatman's certificate is held, a candidate will be required to satisfy the examiner that:
- (a) he understands the general principles of boat management and can carry out orders relating to lifeboat launching and operation and the handling of a boat undersail;
- (b) he is familiar with a lifeboat and its equipment and the starting and running of the engines of a power boat;
- (c) he is familiar with the various methods of launching liferafts and precautions to be taken before and during launching, methods of boarding and survival procedure.

Aboard — In the ship.

About — To go about; to place a boat on the other tack. (*See* "Sailing").

Aft — Towards the stern.

Ahead — Directly in front of the ship.

Ahoy — A word used when hailing.

Alarms — The **fire alarm** aboard ship is the ringing of the bell. The signal for **emergency stations** in all merchant ships is more than 6 short blasts on the whistle or syren followed by 1 long blast. Alarm bells are also fitted in large ships.

A-Lee — To put the helm over to the lee side of the boat.

Alongside — Side by side and touching.

Allotment — After signing-on a ship a seaman may obtain an "Allotment Note" which enables him to allot part of his pay to a relative every month.

Amidships — Halfway between stem and stern, or between the port and starboard side.

Anchors and Cables — Types of anchors:
- (1) The Stockless anchor.
- (2) The Danforth anchor.
- (3) The Admiralty Pattern anchor.
- (4) The Plough anchor.

Stockless Anchor holding in ground.

(1) *The Stockless anchor* — This is the type always carried by ships,

for it has the greater advantage of being self-stowing, *i.e.* the shank is hove right into the hawse-pipe and secured there. The flukes are hinged to the shank and bite into the ground, but if the anchor is **small**, one fluke may dig in more than the other and the anchor will twist and break its hold on the ground. For this reason it is not recommended for small craft.

(2) *The Danforth anchor* has a stock at the crown which projects beyond the flukes, and this ensures that both flukes lie flat and bite into the ground.

(3) *The Admiralty Pattern or Fisherman's anchor* — When the ship is at anchor the stock lies flat on the ground and the lower fluke digs in as soon as a pull comes on the cable, and the anchor holds fast. One disadvantage of this type is that, as one fluke is always sticking up, when the vessel swings to her anchor at the turn of the tide, the cable may foul this fluke and the anchor be torn out of the ground when the weight again comes on the cable. Another disadvantage is that, if the anchor is large it is awkward to hoist on board when it has to be stowed.

(4) *The Plough, Ploughshare or "CQR" anchor* is the type most often used in yachts. The flukes turn on the shank and dig into the ground when the vessels pulls upon the anchor. It is a very efficient type but when large, more awkward to stow than a stockless anchor.

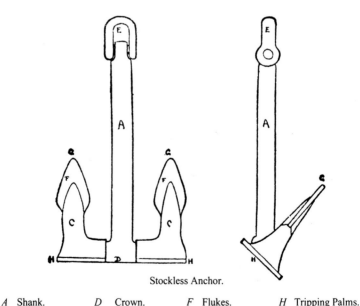

Stockless Anchor.

A Shank.	*D* Crown.	*F* Flukes.	*H* Tripping Palms.
C Arms.	*E* Ring.	*G* Pea or Bill.	

The Flukes of the Anchor are hinged on the Shank. As the Anchor is drawn over the ground the Arms are tripped by the Tripping Palms and Flukes dig into the ground.

The anchors usually carried in ships are (a) **Bower,** (b) **Stream** and (c) **Kedge** anchors.

The two **Bower** anchors are nowadays always stockless anchors, and are the ones normally used in anchoring the ship. A spare bower is always carried. If, in an R.N. ship, the spare bower is stowed in its own hawse pipe and has its own cable it is known as a **Sheet** anchor. The weight of a bower anchor of a ship of 20,000 tons is about $6\frac{1}{2}$ tons.

The **Stream** anchor is usually an Admiralty pattern anchor and is carried in the after part of the ship in case it is necessary to lay out an anchor from the stern. It is not generally provided with a chain cable but is used with a wire hawser.

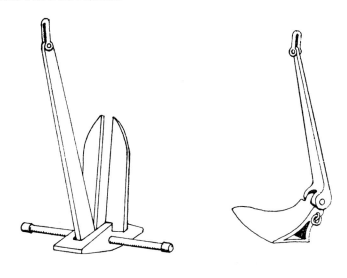

Danforth Anchor.   Plough or CQR Anchor.

The **Kedge** is a small anchor of the Admiralty pattern. It is sometimes used to move small craft through the water by laying out the kedge ahead of the vessel by means of a boat, and then hauling the vessel up to the anchor and repeating the process. This is known as kedging.

Cables — The bower anchors are shackled to long lengths of studded chain cable. The studs strengthen the chain and prevent kinking. The cable is made in lengths of 15 fathoms and the lengths are joined together by shackles. The round end of the shackle is always placed nearest the anchor to prevent the shackle from fouling anything as the cable runs rapidly out. The shackle pin is secured by a peg (usually of wood) driven through the lug of the shackle and the pin. The shackles form a convenient means of knowing how much cable is out. The first shackle out is marked with a wire seizing on the first studded link next to it, the second shackle is marked on the second studded link and so on. Usually the marked links are painted as well.

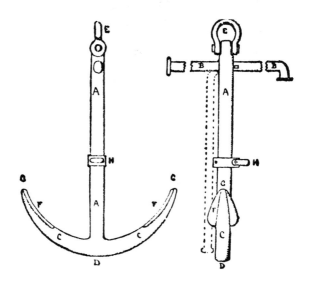

Admiralty Pattern Anchor.

A Shank.	*C* Arms.	*E* Ring.	*G* Pea or Bill.
B Stock.	*D* Crown.	*F* Flukes.	*H* Gravity Head.

The dotted line shows the stowed position of the Stock.
The Stock lies flat on the ground, so that when a strain comes on the cable the lower Flukes dig in and the Anchor holds fast.
The Gravity Band is fitted at the point of which the Anchor balances. It is the place to which the fish-tackle is hooked when the Anchor is being hoisted inboard.

A ship may have 8 shackles — or 120 fathoms — of cable on each anchor.

Chain is the best type of cable, being strong, not easily chafed, and heavy, so that it helps to keep the pull on the anchor horizontal, and thus adds to its holding power. When used by yachts, which usually anchor in shallow water, it should be marked more closely than at 15 fathoms. A simple way is to paint one band of red at 5, one of white at 10, and one of blue at 15 fathoms, then two of red at 20, two of white at 25, and so on.

Smaller yachts may use a nylon warp. This should have a few fathoms of chain between the anchor and the end of the nylon to help the anchor to hold by its added weight and to prevent a rocky bottom from fraying the nylon.

Not less than 25 fathoms or 50 metres should be carried and it is best kept rolled upon a reel. It can be marked by bands of insulating tape thus:

At 5 fathoms or 10 metres 1 band

,, 10   ,,   ,, 20   ,,   2 ,,
,, 15   ,,   ,, 30   ,,   3 ,,
,, 20   ,,   ,, 40   ,,   4 ,,

The **Hawse-pipes** are large steel pipes in the bows through which the cable runs and in which the anchors are secured. The shank of the anchor is hove up into the pipe until the flukes press against the bow plating and the anchor is then secured by a chain rove through the anchor ring and tautened by a rigging screw, Blake Screw Slip, or a claw.

From the hawse-pipes the cables lead to **Compressors** which can be used to grip them, and thence to the windlass.

The **Windlass** consists of a horizontal shaft, turned by steam or electricity, upon which are mounted two head-drums for heaving hawsers, and two **Gypsies**, drums specially shaped to take the cables. The cables pass over the gypsies and down through the **Spurling Gates** to the **Chain Locker** below, in which the cable is stowed and its end secured. The gypsies are free to revolve and are controlled by a powerful band brake. When the anchor is to be weighed the gypsy is connected to the shaft by a screw clutch.

In warships anchors are often weighed by *Capstans*. The cables are led through the hawse-pipes to **Cable-holders**, which resemble gypsies but are mounted flat on the deck. They are driven by the capstan engine which is placed between decks for protection. After passing round the snugs of the cable-holders the cable disappears down the **Navel-pipe** to the chain, or cable-locker. The spurling gates or navel-pipes are provided with plugs and covers to prevent water from getting below.

Blake Screw Slip — A chain stopper fitted with bottle-screw and slip for heaving and securing stockless anchors closely into the hawsepipes. A Blake Slip is similar but has no bottle screw.

Anchoring — Before anchoring, the anchors are cleared away and the ship brought head to wind or tide and stopped. The engines are put astern, and directly the ship has sternway the anchor is let go by releasing the windlass brake (or in the Royal Navy, by knocking off a "Blake slip") as each shackle passes down the hawse-pipe the number of shackles out is signalled to the bridge, either by strokes on the bell, blasts on a whistle, or by showing a numeral flag. When the required amount of cable has been paid out, or veered, the brake is screwed up and when the cable has become taut, showing that the anchor is holding, the ship is said to have **brought up** at her anchor or **to have her cable**. To grip the sea bed the anchor must lie horizontally, so the amount of cable used must be much more than the depth of water: the more cable out the better the anchor will hold. For a short stay in a sheltered anchorage the length of chain should be not less than 3 times the depth of water (at high tide); for a nylon warp, not less than 5 times the depth of water (at high water). In bad weather the other bower may be let go, the ship being given a sheer so that the anchors are spread and the cables lead on each bow and not directly ahead. Sometimes in a strong tide the second anchor is dropped underfoot to prevent the ship from sheering about and thus breaking the anchor out of the ground.

Weighing — To weigh the anchor, the clutch is engaged, the brake unscrewed and the windlass started. Each shackle coming inboard is signalled as before and a hose is played upon the cable to free it from mud. The cable is said to "grow" in the direction it leads from the hawse-pipe. The cable is hove in until it leads "up and down" and the anchor is underfoot. The anchor is then broken out of the ground and hove up into the hawse-pipe and made fast there with the securing chains. Lastly, the spurling gate covers are put on.

Mooring — In places where it is necessary for a ship to occupy as little room as possible, she is moored midway between two anchors. When the first anchor has been let go, twice the usual amount of cable is veered and the second anchor is dropped. The cable on the second anchor is then paid out and that on the first anchor hove in until there is an equal amount on each cable. The ship rides to one anchor and at the turn of the tide swings round almost in her own length and lies to the other. To unmoor, the riding cable is veered and the lee anchor (the one the ship is not riding to) is weighed first.

If a "running moor" is made the first anchor is dropped while the ship still has headway and twice the amount of cable required is paid out; the second anchor is then dropped and cable hove in on the first anchor and paid out on the second anchor until the ship lies between her anchors.

Where there is very little room vessels may have to lie with anchors ahead and astern to prevent them from swinging. In this case the bow anchor is let go as usual and twice the required amount of cable is paid out. Then the second anchor is dropped from the stern and its cable paid out as the forward cable is hove in until the vessel is midway between her anchors.

Laying out an anchor — Yachts often need to lay out a second anchor and may use a dinghy to do so. Make the dinghy fast alongside and lower the anchor over the dinghy's stern, making it fast with a Slip knot (*see* **Rope and Rigging**). Make one end of the warp fast to the anchor (the other end being secured on board) and coil as much of the warp as necessary in the stern of the dinghy. As the dinghy is rowed away the warp is allowed to pay out, and when in position the anchor is dropped. The slack of the warp is then hove in aboard the yacht.

Clearing Hawse — It sometimes happens that after swinging, the cables get twisted together, which is known as a foul hawse. This has to be cleared before the anchor can be weighed. To clear hawse, a stout wire is shackled to the lee cable as low down as possible, hove taut and belayed to the bitts. The cables are then lashed together below the turns. Another wire is led round the riding cable, in the opposite direction to the turn, and shackled on to the lee cable. The lee cable is now unshackled and the turn hove out. After all the turns have been taken out in this way the cable is shackled together again and the lashing cut.

Anchor Buoy — A small buoy secured to the crown of an anchor by a line long enough to reach the surface when a vessel is at anchor. It is thrown overboard ("streamed") just before letting the anchor go. The buoy marks the position of the anchor and is very useful in clearing the anchor if it becomes foul of anything on the bottom. By heaving on the buoy line the flukes of the anchor will be lifted first and should be pulled clear of the obstruction.

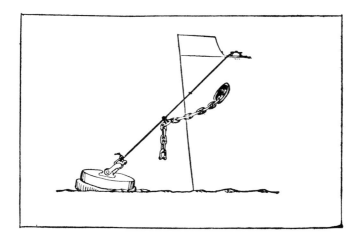

Securing to a Buoy.

Securing to the Mooring Buoy — In some harbours ships make fast to a mooring buoy instead of anchoring. As the cable is used for this, one anchor must first be unshackled, and it is usually hung over the bow secured to the bitts by a stout wire, leaving the hawse-pipe clear for the cable. A large shackle is put on the fifth link of the cable and a mooring wire rove through it. The wire is paid out, a boat takes the end and shackles it to the mooring buoy, and the wire is then hove taut to hold the ship to the buoy. The cable is now "walked back" and slides down the wire. After the cable is shackled to the buoy the wire is slackened until the weight of the ship is taken by the cable.

In harbours where there is no room to swing vessels are moored bow and stern to buoys.

Articles — The "Articles of Agreement" which a merchant seaman signs before joining and when leaving a ship.

Aritificial Respiration — To restore a person apparently drowned. *Immediately*, even while still in the water, place him on his back, hold his head in both hands, one hand pressing the head backwards and the other pushing the lower jaw upwards and forwards. Open your mouth wide, take a deep breath and seal your lips round his mouth while obstructing the nostrils with your cheek. Blow steadily into the lungs and watch for the chest to rise, then remove your mouth. Inflation should be at the rate of 10 per minute. The first 6 should be given as quickly as possible.

Astern — Directly behind a ship.

Athwart — Across; from side to side; transversely.

Avast — Hold fast; stop. *E.g.* "Avast heaving"; " 'vast heaving".

Awash — Level with the surface of the water.

Aweigh — The anchor is aweigh when it is broken out of the ground and off the bottom.

Awnings — Must be slackened in rainy weather as canvas shrinks when wet.

B

"B's for the Bowsprit which points o'er the bow".

Back — The wind is said to back when it alters its direction against the hands of a watch, *e.g.,* from south to south-east.

Backspring — A hawser, usually of wire, used in mooring the ship, put out forward and leading aft, or put out aft and leading forward.

Ballast — Weight put in a ship or boat to help to keep her upright. Nowadays sea water is generally used for ballast.

Barge — A ship's boat used by an admiral.

Barometer Reading 29·53 Inches.

Barometer — An instrument for measuring the pressure of the atmosphere, and so indicating changes of weather. The faster the barometer falls the stronger will be the wind. In the British Isles a S.W. wind with a falling barometer means rain. Barometers are usually graduated in inches (28 to 31 inches). Each inch is divided into tenths and each tenth into five (*i.e.* 0·02 of an inch). Read the inches first, then the number of tenths, and lastly the smallest division, *e.g.,* if the hand pointed to the first small division after 30 inches the reading would be 30·02 inches. (Fig. 1). A barometer may be graduated in units of pressure: millibars. 1000 millibars (1 bar) equals 29·53 inches which is about the average pressure in the British Isles.

Barrico — A small cask, a keg.

Batten Down — To secure the hatch tarpaulins by means of iron battens and wedges.

Beam — The width of the ship at her widest part.

Bear Away — To keep further away from the wind.

Bearing — The direction of anything.

Beating — Sailing to windward.

Becket — A short rope securing-line or handle, *e.g.,* the rope handle of a bucket.

Before — Nearer the bow than, in front of, *e.g.,* before the mast.

Belay — To secure a rope to a cleat or belaying pin. Take a round turn round the cleat with the rope first, and then make figures-of-eight over it.

Bells — The bell is rung in a ship (1) as a fog signal when the ship is at anchor; (2) as a fire alarm; (3) on a passenger ship to warn visitors to go ashore.

It is struck to denote the time every half-hour, starting afresh at each change of the watch.

	Middle Watch				Afternoon Watch	
A.M.	Hours	Bells		P.M.	Hours	Bells
Midnight	0000	8		Noon	1200	8
12.30	0030	1		12.30	1230	1
1.0	0100	2		1.0	1300	2
1.30	0130	3		1.30	1330	3
2.0	0200	4		2.0	1400	4
2.30	0230	5		2.30	1430	5
3.0	0300	6		3.0	1500	6
3.30	0330	7		3.30	1530	7
4.0	0400	8		4.0	1600	8
	Morning Watch				1st Dog Watch	
4.30	0430	1		4.30	1630	1
5.0	0500	2		5.0	1700	2
5.30	0530	3		5.30	1730	3
6.0	0600	4		6.0	1800	4
6.30	0630	5			Last Dog Watch	
7.0	0700	6		6.30	1830	1
7.30	0730	7		7.0	1900	2
8.0	0800	8		7.30	1930	3
				8.0	2000	8
	Forenoon Watch				First Watch	
A.M.	Hours	Bells		P.M.	Hours	Bells
8.0	0800	8		8.0	2000	8
8.30	0830	1		8.30	2030	1
9.0	0900	2		9.0	2100	2
9.30	0930	3		9.30	2130	3
10.00	1000	4		10.00	2200	4
10.30	1030	5		10.30	2230	5
11.00	1100	6		11.0	2300	6
11.30	1130	7		11.30	2330	7
12.0	1200	8		12.0	0000	8

One bell is also struck a quarter of an hour before the change of the watch for the watch below to be called. Seven bells are usually struck at 0720 and 1120; for the watch below to get their meals (*see* "Watch"). Sixteen bells are struck at 0000 on January 1st — eight for the old year and eight for the new.

Bend — To fasten a rope to anything.

Bends and Hitches — The hitches shown in Figs. 1, 2 and 3 have been drawn loosely made so as to show how they are formed.

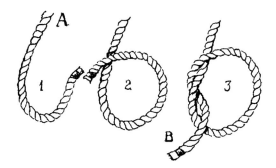

Fig. 1 — Bends and Hitches.

1. A bight. 'A' the Standing Part, 'B' the End.
2. A round turn.
3. An over hand knot.
When practising, make these as shown here.

(1) *Figure of Eight Knot*—Used to prevent a rope unreeving through a block.

(2) *Timber Hitch* — Used to secure a rope to a light case, bale or spar, and also with a *half-hitch* (shown at the top of the figure) for towing a spar.

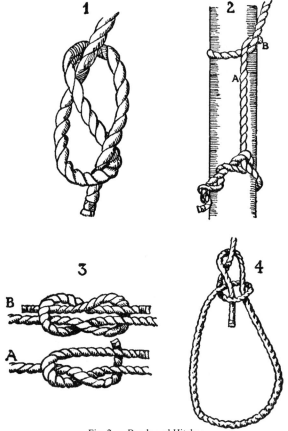

Fig. 2 — Bends and Hitches.

(3) *Reef Knot*—Used for tying two ropes together. Note that both ends finish on the *same* side of the standing parts.

(4) *Bowline* — Used for making a loop on a rope's end.

  (1) Make a round turn where the knot is to be.

  (2) Pass the end up through the round turn, under the standing part and down through the round turn. "The rabbit comes *up* the hole, goes *round* the tree and back *down* the hole".

Two bowlines can be used to join two hawsers together.

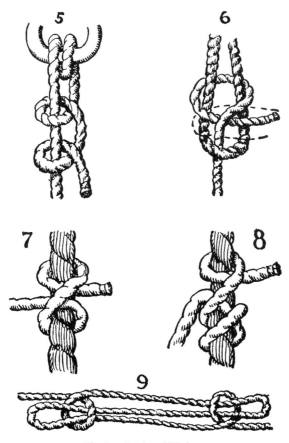

Fig. 3 — Bends and Hitches.

Bowline on a Bight — To lower a man over the side, a double loop can be formed; one being placed under his arms and the other beneath his thighs.

(1) Double the rope for twice the length of the loop required.

(2) Begin to make a simple bowline with the doubled rope.

(3) Instead of tucking the end down through the round turn, pass it over both large loops. Pull these through and work the knot taut.

(5) *Round Turn and Two Half-Hitches* — For securing a rope to a ring.

(6) *Sheet Bend* — For securing a rope to the eye of another. If the end is passed round again, as shown by the dotted line, it becomes a *double sheet bend* and is less likely to jam and will hold better.

(7) *Clove Hitch* — For making one rope fast to a larger one, or to a rail or post.

(8) *Rolling Hitch* — This hitch will not slip down the rope to which it is made fast. The second turn is passed the same side of the standing part as the first turn and rides over it.

(9) *Sheepshank* — For temporarily shortening rope.

(10) *Double Blackwall Hitch* — For securing a rope to a hook. Place the bight of the rope over the strop, cross it behind, place the end over the hook and cross the other part over it. (*See* Figure 2, page 111).

(11) *Marline Hitch* — For securing chafing-gear to a rope, lashing up a hammock, etc. Made like the half-hitch shown in Figs. 5 except that the rope's end is placed *over A* and then under *B*. .

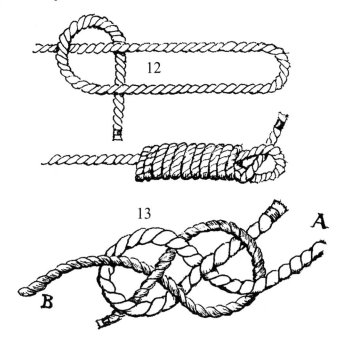

(12) *Heaving Line Knot* — Used to weight the end of a line which is to be thrown.

Form a bight, and bind the two parts together until the end is almost reached. Pass the end through the loop left and then pull taut the standing part.

The Heaving line knot can be used to join nylon or gut and to join nylon or gut to a fishing hook.

Joining nylon or gut: make a heaving line knot at the end of one length, pass the end of the other length through the eye and then make a second heaving line knot with it. Then pull both knots taut.

To bend on a hook: pass the end of the line through the eye and along the shank of the hook. Make a heaving line knot winding the line round both its parts and the shank of the hook.

Nylon must be well moistened, and gut soaked, before manipulating them. (Figure, Page 16).

Nylon rope is smooth and slippery and some of the standard bends and hitches tend to slip.

(13) *Carrick Bend* — Used to join two hawsers together. It can be used equally well for ropes of different sizes.

    (1) Form a loop at the end of rope A (the larger, if they are not of equal size).

    (2) Lay rope B over this loop, pass it *under* the standing part of A, *over* the end of A, *under* the left part of the loop, *over* B's standing part, and *under* the right part of the loop. The two ends should be on opposite sides of the two standing parts. The ends may be stopped to the standing parts for additional security. (Figure, Page 16).

Slip knots which can readily be released are occasionally useful.

(14) *Highwayman's Hitch* — Pass the rope through a ring or round a post and make an Overhand knot round the standing part, but without pulling the end through. The knot can therefore be undone simply by pulling the end. A Clove hitch and Sheetbend can be similarly used as Slip knots if the bight, instead of the end, is pulled through.

Bight — A loop of rope.

Bitts — Pairs of short stout posts to which ropes may be made fast.

Blue Peter — Flag P of the International Code.

Boats and Boatwork — Types of ship's Boats.

In the Royal Navy the following kinds of boats are used:

Motor Launches — 34 feet long; twin engines, fast, manoeuvrable, and good in rough water.

Motor Boats — 24 feet long, speed about 10 knots.

Motor Whaler — 27 feet long, double-ended fitted with an air-cooled engine and, usually, for sailing also. Good seaboats and manageable in surf for beaching.

Inflatables — About 16 feet long, of rubber and wood construction, and with outboard engines. Much better seaboats than they might appear to be.

Rigid Inflatable Boat — R.I.B. A rigid G.R.P. hull bottom with inflated tubes forming the sides. Transom usually wood to take an outboard engine.

Barges — Especially smart, fast boats reserved for an Admiral's use.

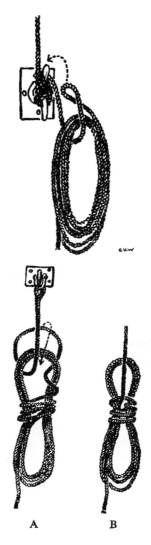

A   B

Two ways of hanging up a coil of rope.

Cutters — 32 feet long; pulling 12 oars, double-banked, sloop rigged, are used for training purposes.

Merchant Ships carry:

Lifeboats — Of varying lengths from 24 to 40 feet (the dimensions and the number of persons which can be carried are cut upon, or near, the

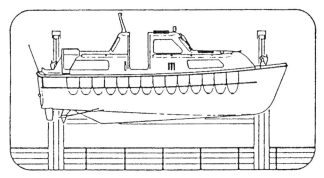

12·4m Partially Enclosed Lifeboat/Launch — 106 Persons.

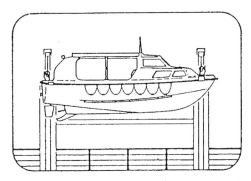

8·5m Partially Enclosed Lifeboat/Rescue Boat/Radio Cabin.

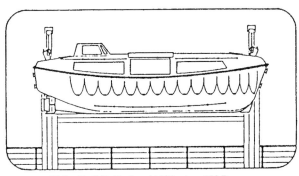

11·0m Partially Enclosed Lifeboat — 150 Persons.

stem); double-ended; fitted with buoyancy which will keep the boat afloat if she is swamped. Lifeboats are either totally or partially enclosed and fitted with diesel engines.

Foreign-going ships carry at least one motor lifeboat, fitted with a diesel engine and capable of a speed of 6 knots when loaded, with sufficient fuel for 24 hours; a radio installation, a searchlight with a lamp of 80 watts. Motor lifeboats are capable of towing liferafts.

Passenger ships must carry sufficient boats to hold all their passengers and crew. Cargo ships must carry enough boats *on each side* to do this, or one boat at the stern.

Merchant ships' lifeboats are always numbered, odd numbers on the starboard side, even numbers on the port side. If one set of davits handles a second boat this boat has the letter "A" added to the number. The name and port of registry of the ship to which the boat belongs is painted on each side of the bow.

Yachts carry:

Dinghies — Inflatable or otherwise. A pram, or pram-dinghy has a small square transom at the bow as well as the usual larger one at the stern.

Sea-Boats, or Accident, or Emergency Boats — In warships and passenger ships, when at sea, one boat on each side of the ship is kept uncovered, with the plug in, ready for immediate use in case of "man overboard". They are fitted with a boat rope led well forward which is let go from the boat after she is afloat. This is better than using the boat's painter which would have to be hauled into the boat. Whalers are used as sea-boats in warships, and lifeboats of 26 feet or less in merchant ships. Cargo ships with only one lifeboat may carry an R.I.B. for man overboard use.

Boat Design and Construction — Boats may be built of steel, aluminium alloys, glass-reinforced plastic, or wood, either planks or marine plywood.

Most boats are now built of **Glass Reinforced Plastic**, GRP, which consists of layers of glass-fibre impregnated with polyester resin. A number of these layers are laid over a mould and form a laminate. The whole hull bonds together in a very strong construction which needs practically no maintenance.

Wooden Boats — A boat's planking may be arranged in one of three ways: *viz.*, Boats are said to be (1) **clinker** or **clencher** built when the planks run fore and aft and overlap each other like the tiles of a roof; (2) **carvel** built when the planks run fore and aft and do not overlap each other but make a smooth surface; (3) **diagonal** built when the planking is doubled and the planks of the outer skin run diagonally across the planks of the inner.

Ship's boats up to and including cutters and moderate size lifeboats are clinker built, a system which combines lightness and strength. Larger boats are diagonal built for strength.

The planking of a boat is strengthened by and fastened to a skeleton or framework (Fig. 4). The keel forms the backbone of the boat and it is strengthened inside the boat by the **hogpiece** (to which the lowest plank known as the **garboard strake** is fastened) and the **keelson** (pronounced kelson). The ends of the keel are joined to the **stempost** and **sternpost**, the joints being strengthened by the **deadwood**. Strengthening the top of the planking and fastened to the top plank (which is called the **top strake** or **sheer strake**), is the **gunwale** (pronounced gunnel), and this is protected by a thin **gunwale capping**. The crutches ship in the gunwale. Cutters and larger boats have a **washstrake** fitted on top of the gunwale, openings called **rowlocks** are left for the oars, and these are filled in by wooden **poppets** when the boat is not being pulled. The side of the boat is protected by a stout **rubber** or **rubbing strake**. Inside the boat, running from gunwale to gunwale and fitted between hogpiece and keelson, are the **timbers** or **floors** forming the ribs of the boat.

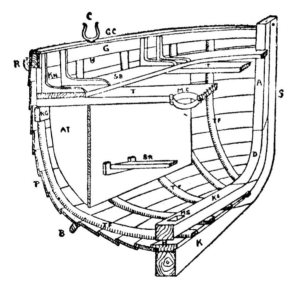

Fig. 4 — Clinker Built Lifeboat.

A	Apron.	*H*	Hog.	*Z*	Rubber.
AT	Airtank.	*K*	Keel.	*ZG*	Rising.
B	Bilge piece.	*KS*	Keelson.	*S*	Stem.
C	Crutch or row-lock.	*KN*	Knees.	*SR*	Stretcher.
D	Deadwood.	*MC*	Mast clamp.	*SB*	Side bench.
G	Gunwale.	*MS*	Mast step.	*T*	Thwart.
GC	Gunwale capping.	*P*	Planks.	*TF*	Timber or floors.

A hook, by which the boat is hoisted, is fitted at each end of the boat being bolted through both keelson and keel. It is not shown in the figure for the sake of clearness. Royal Navy boats are not fitted with hooks but are provided with chain slings.

The iron stem band which protects the fore part of the stern is omitted for the same reason.

Strengthening the boat athwartships are the **thwarts** on which the rowers sit. The thwart ends rest upon the **rising** and are secured to the sides of the boat by **knees**. In lifeboats, **side benches** are fitted on top of the thwarts, and underneath these are either airtight buoyancy **aircases** made of yellow metal or cases of **buoyant material** occupying one-tenth of the space inside the boat. Boats which have no air tanks are only fitted with benches round the sternsheets, the **sternsheets** being that part of the boat abaft the after thwart. The **headsheets** is the small platform in the bows of the boat. A **backboard** is fitted in some boats across the back of the stern benches. The **transom** is the flat wood which runs athwartships from the sternpost to the ends of the planking on both sides in square-sterned boats; it is not found in whalers and lifeboats. The piece of wood fitted on the after side of the stempost (and on the fore side of the sternpost in lifeboats and whalers) is known as the **apron** and helps to secure the ends of the planking to the post. The bottom of the boat is protected inside by **bottom boards** or gratings, and **stretchers** are fitted across the boat on which the rowers brace their feet. Some boats are fitted with **bilge pieces** on the turn of the bilge (the bilge may be described as that part of the boat where the bottom becomes the side); they have handholes cut in them, useful if the boat capsizes. When the bottom and the side of the boat meet sharply at an angle the boat is said to be **hard-chined**. Some boats are provided with **dropkeels** or **centreboards**, metal fins lowered from inside the boat through a slot in the keel. This prevents the boat from making leeway when sailing close-hauled, and when the boat is not under sail it is hauled up inside its casing. The **rudder** is slung by **pintles** which ship into **gudgeons** fixed in the sternposts, or braces on the rudder ship over a rudder-bar fixed to the sternpost in such a way that the rudder cannot accidentally become unshipped. It is moved by a **tiller** or by a **yoke** and **yoke lines**. (*See* page 191). A projection of the keel aft beneath the propeller is known as a **skeg**.

BOAT PULLING

The most important thing to remember in boat pulling is to **keep together**. The boat's crew first seat themselves on the thwarts facing aft, ship their crutches and grasping their oars by the looms **Point the oars**, *i.e.,* lay the blades on the gunwale ready for tossing, except the bow and stroke oarsmen (the foremost and aftermost men in the boat) who keep the boat alongside with their boathooks, waiting for the order to shove off.

At the order **"Toss your oars"**, the oars are tossed up on end together, the looms resting on the bottom boards and the blades feathered fore and aft.

At the order **"Oars down"** the oars are lowered into the crutches and brought to the "Oars" position, shafts horizontal, blade feathered flat fore and aft.

If the oars are not to be tossed, the order **"Ship your oars"** or **"Oars ready"** is given and the oars are placed in the crutches.

At the order **"Stand by"** the crew lean aft, straighten their arms and turn the blade ready to take their first stroke.

GOOD POSITION.

Fig. 5 — BAD POSITION.

Sitting too near gunwale. Wrong grip on oar. Only one foot on stretcher.
Back too straight. Eyes not in the boat.

At the order **"Give way together"** they start to row, taking time from the stroke oars, the port stroke taking time from the starboard stroke. It is important to keep a good position when pulling. Sit about two-thirds on the thwarts, heels together on the stretcher, back and shoulders straight. Grasp the oar with the hands about 1 foot apart and keep the arms straight during the stroke letting the back and thighs do the work. Keep your eyes on the man in front; don't round the back, reach too far aft, or watch your oar. Between each stroke feather the oar to lessen the wind resistance by dropping the wrists to turn the blade horizontal, but be sure to make the stroke with the blade vertical or you may "catch a crab".

The order **"Oars"** is given to rest the crew, who lie on their oars as described above. If the boat has to be turned short round the order **"Hold water port"** (or starboard) is given. The oars on the side named are then held in the water, the other side continuing to pull. If the order **"Back together"** is used, the oars are pushed backwards through the water. When the boat is round the order **"Up together"** is given and the side which has been holding or backing water pick up the stroke again, taking time from the other side.

Before coming alongside the order **"Bow"** is given, and the bow oarsman tosses his oar and boats it, and stands up facing forward with his boathook tossed ready to fend off; that is, to keep the boat from striking against anything.

"Way enough" is given when coming alongside, followed at once by **"Toss your oars"** when the oars are tossed, or **"Boat your oars"** when the oars are tossed out of the crutches and lowered gently into the boat, blades forward and looms in line with the after thwart, and the crutches unshipped. In a single-banked boat only the bow oar is tossed in; the blades of the rest are allowed to swing aft and the oars boated.

To keep the boat under better control the coxswain may vary his orders by boating first only the oars next to the ship's side. The order might be **"Way enough, boat the starboard oars"**. All would take one more stroke, then the starboard oars would be boated and the port oars left at the oars position.

The orders "Oars", "Bow", and "Way enough", are all carried out *one stroke after the order is given*.

A boat is always brought alongside head to wind or tide, whichever is the stronger.

Dinghy Rowing — The stroke should be as long as possible, but in choppy water may have to be short and quick, with the oars not fully feathered, and the blades lifted higher out of the water than normally.

Sculling a Boat (Fig. 6) — An oar is placed in the notch in the transom (or in a crutch shipped there) and the oar moved from side to side, without the blade being removed from the water, in a figure-of-eight motion. This method of propelling a boat requires some practice, but is a useful accomplishment.

Fig. 6 — Sculling a Boat.

Embarking — When getting into a small boat do not step onto the gunwale (which may list the boat over) but, carefully, as near the centre-line as possible, and sit down at once.

In larger boats, when officers are embarking or disembarking it is customary for the senior to enter the boat *last* and leave it *first*.

BAD WEATHER

If a boat can be kept end-on to the seas she will ride them safely, but if she is allowed to broach-to, that is, to get broadside on, the wave crests may swamp or capsize her.

To keep her end-on a **sea anchor** (Fig. 7) may be used. This is a floating drogue which, being practically submerged, remains stationary in the water. A boat will drift to leeward much faster, so, if the bow of the boat is attached to the sea anchor by a riding line, the bow will be continually pulled to windward.

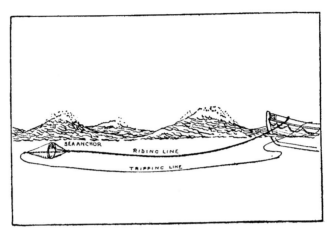

Fig. 7 — Riding to a Sea-anchor.

Sea Anchor — A cone-shaped bag made of porous, slightly stiff material which allows water penetration. The painter must be of rot proof material and braided construction. The painter must be 30 metres long and have a breaking load of 10 kilo Newtons. The sea anchor mouth shall open immediately on deployment.

Dimension for a boat over 9m.

Mouth diameter 800mm.

Sloping length 1050mm.

Shroud lines 1050mm.

When a sea anchor is used it is most important to protect the riding line from chafe where it passes over the gunwale. It should be well parcelled, and the parcelling renewed when necessary, or the painter will soon chafe through and part.

Landing on a Beach — Surf is more dangerous than it appears to be from seaward. The danger lies in the breakers tending to throw the boat broadside on and capsizing her. To prevent this each wave must be met end on and allowed to pass as quickly as possible. As the rudder is no use in surf, before approaching the breakers unship the rudder and ship the steering oar. Tow the sea anchor astern and trip it with its tripping line. As each wave approaches let go the tripping line so that the sea anchor fills and pulls at the stern of the boat, at the same time backing the oars together. Directly the crest of the wave has passed trip the sea anchor and pull inshore until the next wave overtakes the boat. A square-sterned boat should be backed in stern first. (*See* also; **Davits; Handling Craft; Life-saving Appliances; Sails and Sailing**).

Boatswain — The chief petty officer on a merchant ship.

Boatswain's Chair — A seat made from a board secured in a strop, used in work aloft. To make the gantline fast to the chair so that you can lower yourself, push a bight of the line through the eye at the top of the strop and then pass the bight right over the chair and pull taut.

Bollard — A stout post to which ropes are made fast. When placing the eye of a hawser over a bollard which already has another hawser made fast to it, put the eye up through the other hawser's eye before dropping it over the bollard. Either hawser can then be let go without disturbing the other.

Boom — A spar for stretching the foot of a sail. Any long spar or piece of timber.

Boot-topping — The composition used for painting a ship's water-line.

Bottle-screw — A screw used for setting up rigging. It consists of a central part at each end of which a screw eye is fitted. On the central part being turned with a spike both eyes are screwed towards each other this making the rope taut.

Bow — The front, or forepart, of a ship. **On the bow;** in a direction between right ahead and abeam.

Bowse — To pull closely together.

Box the Compass — To name the points of the compass in regular order.

Breast Line — A hawser used to haul or keep a ship close alongside a quay.

Breasthooks — Flat triangular plates joining the bow plating.

Breaker — A small water cask.

British Seaman's Card — Issued to British Merchant Seamen. Useful as a means of identification and as a passport in foreign ports.

Brought-Up — Anchored securely, brought to an anchor.

Brow — A gangway from ship to quay.

Bullrope — A hawser used to keep the bow of a ship from bumping against a mooring buoy.

Bulwark — The ship's side above the deck.

Bumboat — A shore-boat used to carry provisions, etc.

Bunkers — Compartments in which fuel is stored. Also the fuel itself.

Bunting — Material from which flags are made.

Buoys — Used to mark the edges of channels, shoals, etc. To enable one to distinguish their function they are of different shapes and colours and may carry topmarks and lights.

The International Association of Lighthouse Authorities buoyage System A consists of two types of mark.

Lateral marks denote the port and starboard sides of the channel according to a conventional direction of buoyage which is usually in towards a port but which is always marked on the chart.

Can shaped buoys painted red mark the port side. They may have can shaped topmarks and may show red flashing lights.

Cone shaped buoys painted green mark the starboard side. They may have cone shaped topmarks and may show green flashing lights.

Cardinal Marks are named North, South, East and West and indicate that the navigable water lies on the named side of the mark. For example, it is safe to pass to the Eastward of an East mark.

Cardinal marks are pillar buoys painted with black and yellow bands according to a system shown in the diagram. If they have topmarks these consist of two cones which may be points up, points down, points together or bases together as indicated. White flashing or quick flashing lights may be shown. Quick flashing for North, Flashing group three for East, Flashing group six for South and Flashing group nine for West.

Safe water marks mark a landfall or the middle of a wide channel.

Isolated danger marks are placed on dangers which have safe water all round them.

Special marks are used to show areas such as firing practice areas. It is best to consult your chart.

Burgee — A swallow-tailed flag: The distinguishing flag (usually triangular) of a yacht club.

By the Head — A vessel is by the head when she is deeper in the water forward than aft.

By the Wind — Sailing as close as possible to the wind; close-hauled.

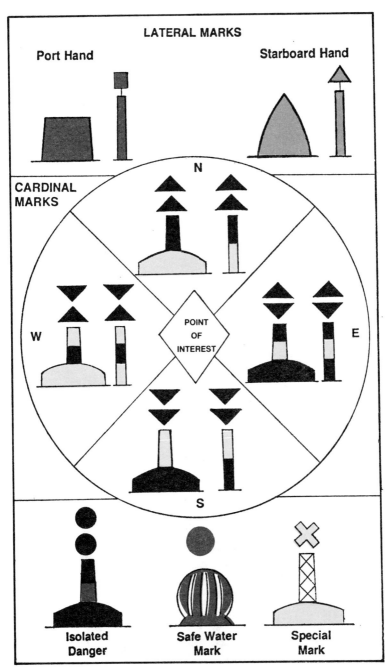

System A Buoys.

C

"C is the Capstan we merrily turn round".

Cab — A shelter at the corner of the bridge.

Cable — A tenth of a nautical mile; approximately 200 yards.

Canhooks — Two hooks on a length of chain used for lifting casks.

Cant — To tilt or incline.

Canvas — Made of flax and supplied in bolts 40 yards long and 2 feet wide. Tarpaulin canvas is waterproof, not so strong and wider. (*See* under "Sail-making" for ways of sewing canvas).

Camber — The arching of a deck which allows the water to run off.

Capsize — If a boat is capsized or swamped the crew should not attempt to swim ashore but should stay with the boat, which should still be able to support her crew.

Cargo — Cargo, once loaded, is in the direct care of the ship and so must be protected from all damage and pilferage.

GENERAL RULES FOR STOWAGE

(1) Cargo must be **securely stowed** so that it will not get adrift at sea when the vessel rolls and pitches.

(2) Cargo must be kept **clear of the deck** by stowing it in on pieces of wooden dunnage. It is kept clear of the ship's side by fixed wooden cargo battens. This prevents damage from any water in the hold or from moisture known as "sweat", which condenses on the ship's side.

(3) Cargo for different ports must be **separated** by dunnage, or clearly marked.

SPECIAL CASES

Bags — Must be well protected from all contact with steel by dunnage mats.

Bales — No hooks should be used with bales of goods or the contents will be torn.

Bulk Cargo — Grain is usually carried in bulk and great care has to be taken to see that the holds are as full as possible so that the cargo does not shift, ships having been capsized by grain shifting to one side. To prevent shifting, a fore and aft bulkhead of **shifting boards** is rigged to each hold and feeders are constructed in the hatchways, so that as the grain settles in the hold any empty space is filled up from the grain in the feeders.

Cases of goods should be stowed with their marks and numbers uppermost and any instructions marked on them (*e.g.,* "This side up") attended to.

Dangerous Chemicals, etc. — Are stowed on deck where they can be inspected for leakage and jettisoned if necessary.

Ventilation — Many cargoes become damaged or dangerous if not properly ventilated. Some require a system of through-ventilation (*e.g.,* rice) others surface ventilation (*e.g.,* coal).

DUTIES OF MATES OF HOLDS

When acting as hatchman or mate of hold bear in mind the following:

Always be at the hatch when it is opened, and see that hatches and tarpaulins are put on properly when closed.

Never leave the hold whilst stevedores are below.

Keep a continual watch to prevent pilfering or damage; if you find any pilfered or damaged cargo report by note to an officer, leaving the package in the position in which it was found.

Do not allow frail packages to be put in the same sling as heavy ones, and see that the cargo is reasonably handled at all times.

See that slings and snotters are in good condition, condemn any that are worn and chafed.

Take great care that the cargo is put out at the port for which it is marked.

When the cargo has all been discharged for the port you are in, make a careful search to see nothing is over-carried.

Note the time of opening and closing the hatch each day.

Before taking in cargo see that the deck is dry, clean, and well dunnaged.

Carry out any instructions as to where the cargo is to be stowed, and on no account let it be stowed otherwise.

Don't stand under any derricks or open hatchways. Keep your eyes open.

Hatchways are closed by (1) steel hatch covers or (2) by movable beams shipping inside the hatch coamings. Wooden hatches are then placed over the beams and on top of these two or more tarpaulins are spread (oldest ones on top). These are secured by long steel battens placed in cleats riveted outside the hatch coamings and kept in position by wooden wedges hammered home. If only half the hatch has been uncovered while working cargo the hatch beams still in position should be secured by bolts to ensure that they are not accidentally unshipped.

Tallying — At the top of the page in the tally book write the port you are in, and the date. Underneath that, the kinds of goods being loaded or discharged, their marks, and the port to which they are addressed. Then the number in each sling, or the number carried in or out.

Example of tallyings
LONDON   23rd May, 1974.
c/s (cases) Whisky S. H. & Co.   DURBAN.
‖‖‖ ‖‖‖ ‖‖‖ ‖‖‖ ‖‖‖   25
‖‖‖ ‖‖‖ ‖‖‖ ‖‖‖   18
                                          ——
                                          43
                                          ——

Stowage Factor — The number of cubic feet occupied, on the average by 1 ton weight of cargo allowing for packing, dunnage, etc.
Examples:
  Bags — cement 35, sugar 42, flour 45;
  Bales — cotton 80;
  Bulk — ores 12 to 20, coal 40 to 45, wheat 47.
If expressed metrically: cubic metres per tonne.
Carry Away — To break.
Carry On — To proceed.
Cat, or Una, Rig — A single mast, stepped right forward, with a gaff or Bermuda sail.
Catamaran — A vessel with two hulls. (2) A flat wooden float used in docks as a fender etc.
Cat's Paw — A light air, distinguished by ripples, made on calm water.
Caulk — Wooden decks are made watertight by being caulked with oakum which is hammered in between the planks and covered with pitch.
Centre-castle — The raised part of a ship's hull amidships.
Chart — A sea-map showing depths of water, etc.
Cleat — A piece of wood or metal used for belaying ropes.
Close-hauled — Sailing close to the wind.
Coamings — The edges of a hatchway.
Collier — A vessel employed in carrying coal cargoes.
Collision — The master of each vessel is required to give whatever assistance he can to the other vessel and to stand by her until assistance is no longer required. He must give the name, port of registry of his ship, where she has sailed from and where she is bound. A full entry must be made in the log. On arrival in a British port a report of the incident should be made to the Customs Officer.
Column — A derrick-post.
Come-up — After hauling a rope, to come up is to light or hand the rope forward so that it can be made fast.
Compass — The compass is an instrument which indicates direction. It enables ships to be steered in any direction that is required, and also shows the direction of any visible object.
Compasses are of two kinds:
  (1) The Mariners' or Magnetic Compass.
  (2) The Gyro Compass.

The Magnetic Compass consists of two or more magnetised **needles** attached to a **card** on which the points and degrees are marked, balanced on a pivot and contained in a **compass bowl**. Most ships and all boats have **liquid compasses**, in which the compass bowl is filled with a mixture of alcohol and water (the alcohol preventing the water from freezing), so that the compass card partly floats. Liquid compasses often have smaller cards than dry compasses to reduce the effect of eddies set up when the card swings. The liquid makes the card very steady and less affected by violent rolling, vibration or gunfire than dry compasses.

The compass is slung in a stand called the **binnacle** by means of **gimbals**, rings which permit the compass bowl to remain level however unsteady the ship and binnacle may be. The binnacle is lit by electricity, and the binnacle top may be fitted with an oil light in case of electrical failure.

The magnetic compass does not point to the true north pole, but to a locality in the far north of Canada called the magnetic north pole. The difference in direction between these two poles is called the **variation**.

Ships, being built of steel, deflect the compass from pointing to the magnetic pole, the error thus caused being known as the **deviation**. The deviation is greatly lessened by fitting the binnacle with magnets, two iron spheres and an iron Flinders bar, which are so placed to counteract the attraction of the ship. As all iron and steel deflect the compass needles, it is important that no knives, iron buckets, etc., should be near the compass. The combined variation and deviation is called the **compass error** and must be allowed for when shaping a course.

The principal compass in a ship is mounted on the bridge where there is a good all round view and is called the **standard compass**. Other compasses are placed at the steering positions.

The standard compass is provided with an **azimuth mirror** which fits on the top of the compass bowl and can be turned in any direction. It is used to find the direction, or bearing, of any object, and has a prism in which the observer can see a reflection of the compass card as he looks at the object.

When removing the binnacle top be careful not to sweep off the azimuth mirror at the same time!

Grid Compasses — A grid steering compass has a top glass cover marked with a "grid", which may be a pair of parallel lines or an arrow. The glass can be rotated and its edge is graduated.

To steer a given course, the glass cover is first rotated until the required course is opposite the **lubberline**, the line in the bowl fixed in the exact direction of the ship's head. The vessel is then steered so that the N-S line on the glass cover is over and exactly parallel with the N-S line on the compass card.

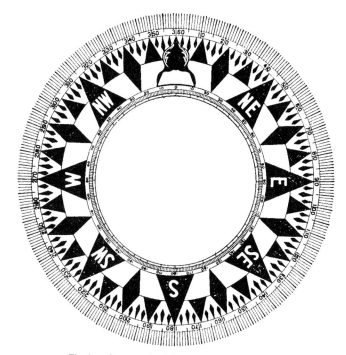

Fig. 1 — Compass Card showing Points and Degrees.

The Gyro Compass — Almost all ships and some big yachts carry gyro compasses in addition to magnetic compasses.

The gyro compass has nothing to do with magnetism, but is controlled by the earth's rotation. It consists of a wheel turned very rapidly by an electric motor. The axle of the wheel points to the true north and south poles (*not* the magnetic ones) for the reason that a rapidly spinning wheel, free to turn in any direction, can be made to keep its axis parallel to the earth's axis.

The master compass, which consists of the wheel and motor, may be placed between decks for protection; and repeater compasses, which are driven electrically from the master compass, are mounted on the bridge and wherever they may be required.

BOXING THE COMPASS

The compass card is divided into points and quarter points and also into degrees. (Figs. 1 and 2).

The Points — The four principal or cardinal points, north, east, south, and west, divide the compass card into four quadrants or quarters. Half way between the cardinal points are the half cardinal or quadrantal points, north-east, south-east, south-west and north-west. Half way between the cardinal and half-cardinal points are the intermediate or three-letter points, which take their names from the cardinal and half-cardinal points between which they lie, the cardinal points being named first; north-north-east, ENE, ESE, SSE, SSW, WSW, WNW, and NNW. Between all the points already mentioned are sixteen "by" points which take their names from the near-by cardinal and half-cardinal points. So that we get thirty-two points, eight on each quadrant, each quadrant being divided in the same way.

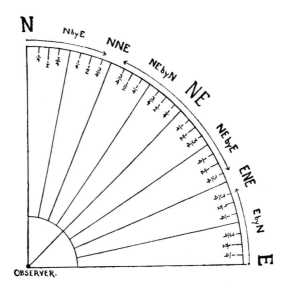

Fig. 2 — N.E. Quadrant of Compass showing $\frac{1}{4}$ and $\frac{1}{2}$ Points.

The points in the first quadrant are:

The cardinal point	North
The point next to it	N by E
The three-letter point halfway between N and NE	NNE
The by point next to the half-cardinal	NE by N
The half-cardinal	NE
The by point	NE by E
The three-letter point between E and NE	ENE
The by point	E by N
The cardinal point	E

The card is still further divided into quarter and half points, which, like the "by" points, are all named from the nearest important points, *i.e.,* the cardinal and half-cardinals (*not* from the three-letter points).

When boxing the compass (as naming the points in their correct order is called), always think of yourself as being placed in the middle of the card, as in the figure.

Half and quarter points are now little used, but a seaman should be able to box the compass in **points** and understand the arrangement of the **degrees**.

The Degree — Boats' compasses are only marked in points and half points but all others are divided into degrees, and it is in degrees that the course is usually given and steered. The card is divided into 360 degrees, *i.e.,* 90° in each quadrant. Most compass cards are marked right round from the north through 90° (east), 180° (south), 270° (west) and 360° (north). In some magnetic compasses each quadrant is marked from 0° to 90° beginning at north and south, so that (for example) SE on a gyro card is called 135° and on some magnetic compasses, S45°E. NE or N45°E is written 045° when the 3-figure notation is used, and spoken "zero four-five".

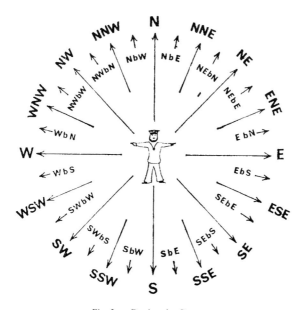

Fig. 3 — Boxing the Compass.

Container — A rectangular container constructed to withstand the various forces placed upon it by the weight of cargo it contains, the weight of other containers stacked upon it and the motion of the ship at sea. The advantages of this method of carrying cargo are that it can be loaded at source, sealed and unloaded at its final destination without intermediate handling of the contents. This helps to prevent handling damage and pilfering. The amount of cargo in a container, typically forty feet long, eight feet wide and eight feet high, with a payload of 50,000 pounds, means that with specialised handling equipment the turn round time is very short.

Convoy — A number of merchant ships in company and under the orders of a senior officer.

Counter — The overhanging part of the stern.

Course — The direction in which a ship is going.

Coxswain — The man in charge of a boat.

Cranky — A vessel easily listed or heeled over.

Cringle — A horse-shoe shaped grommet fitted with a thimble worked into the bolt-rope of a sail or awning and used in securing it.

Current — Water flowing constantly in one direction (unlike tides which change and flow in the opposite way). Named from the direction *towards* which it flows.

Cyclone — Winds blowing towards and round an area of low pressure. In the China Seas a violent storm of this description is called a typhoon; in the West Indies a hurricane.

D

"D's for the Davits to lower boats down".

Davits — Davits are used to hoist boats in and out of the water. They are placed in pairs at the ship's side, one davit at each end of the boat. There are three classes of davits.

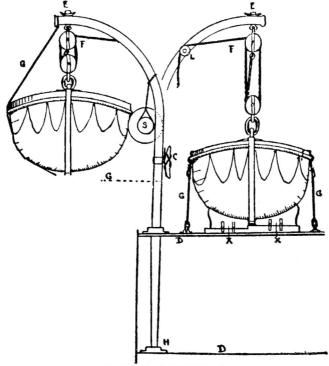

Fig. 1 — Davit (Radial Type).

The figure shows the boat turned in and resting in the chocks, and also swung out and secured against the griping spar.

C   Cleat for boat's fall.
D   Deck.
E   Eye-plate to which is shackled the span and guy.
F   Boat's fall.
G   Gripes, fitted with slips to enable them to be readily cast off.
H   Step or socket for heel.
S   Griping spar, fitted with puddings against which the boat is secured.
X   Chocks.

Radial Davits — These (Fig. 1) are made to turn round about. The davits heads are controlled by guys. The boat in its stowed position rests in chocks on the deck secured there by gripes. To swing the boat out ready for lowering (1) see that the falls (the tackles by which the boat is hoisted) are taut and well fast; (2) knock off the outside gripes and wait for the men to get inboard again; (3) down chocks; (4) knock off inside gripes. The boat is now free to swing out. Haul the after guy and as the boat swings aft push out the bow, then haul on the forward guy and as she goes forward push the stern out, then square up the davits and make the guys fast.

Boats may be kept in the swung out position. In this case a long spar called a griping spar is a slung between the davits level with the boat's rubber and the boat is secured against it by long gripes leading from the davit heads, passing round the outside of the boat and making fast on the opposite davit. When the gripes are cast off the boat is ready for lowering.

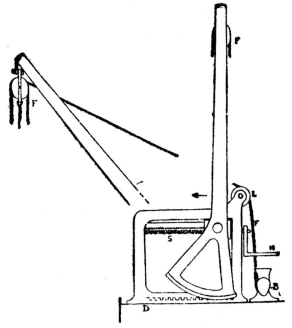

Fig. 2 — Davit (Luffing Type).

B Lowering bollard.	*F* Part of boat's fall.	*L* Fairlead.
D Deck.	*H* Handle.	*S* Screw.

Luffing Davits — These Davits (Fig. 2) are placed at the extreme ends of the boat and do not turn round but are moved by a screw motion as shown in the figure. To swing out: (1) see that the falls are taut and well fast; (2) knock off outside gripes and wait for the men to get inboard; (3) down chocks; (4) knock off inside gripes; (5) ship the handles and turn the screw gear.

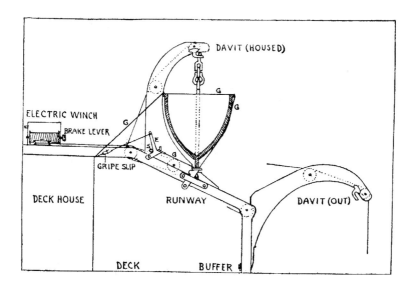

Fig. 3 — Davit (Gravity Type).

In the Stowed position — The boat is secured to the davit by the gripes (G). The gripes lead over the top of the boat, down and under the keel and the other ends ship over the short arm of the gripe elbow (E). The gripe elbow bears against the stud (S), which projects from the davit and thus prevents the davit from sliding down the runway. The gripe elbow is held in position by the other end of the gripe which is spliced into the long arm of the gripe elbow and held by a slip. When the slip is knocked off, the gripe elbow falls outward, leaving the boat free in the davit, and the davit free to move down the runway as soon as the brake is released.

The wire falls lead from the davit head over a fairlead to a sheave at the top of the runway then round a sheave at the foot of the davit, back again over another sheave at the top of the runway, and then inboard to the winch. **In the swung out position**, the toe of the davit bears against a buffer.

The davits and skids are made of steel plates with the sheaves placed between them.

Gravity Davits — These (Fig. 3) are made in two principal parts, the davits proper being mounted on skids above the decks. The weight of the boat is made to do the work of swinging out, and the wire falls are led to a winch, placed between the skids, which is fitted with one barrel for each fall, both barrels being controlled by one brake lever. The boat and davits stow on the sloping runway of the skids, being secured there by gripes. To swing out the boat; knock down the centre keel support and knock off the gripes. Release the brake and the davits and boats will slide down the runway to the swung out position. The davits stop there and, if the brake is kept off, the boat continues to lower into the water.

Lowering — Where manilla rope falls are fitted the falls are led from the davit head to a stout lowering bollard, staghorn or cleat and the rest of the fall is reeled up on a reel. The boat lowerers stand or sit firmly behind the lowering bollard, and remembering that a boat may weigh as much as 3 tons, lower her carefully on an even keel. Where wire falls are fitted they are led to a winch and lowered as described in the previous paragraph. Lifelines are fitted to hang from the span which runs between the davit heads for the boat's crew to hang onto. If the ship is rolling or pitching, frapping lines should be passed from the ship round the falls and back again to keep the boat as steady as possible while she is being lowered.

When the boat is in the water unhook the after fall first, for if the forward one is unhooked first the boat's bow may sheer out from the ship's side and it will be difficult to unhook the after fall. Then, if the painter is used, haul on it and put the tiller over towards the ship. This will sheer the boat away from the ship's side.

Disengaging gear may be fitted which will release both falls at the same instant.

Hoisting — See that there are no turns in the falls and hook on the forward fall first. When manilla falls are fitted they are led through snatch blocks to a winch, or if the boat is to be hoisted by hand the falls are led along the deck and hauled taut singly, then married (*i.e.,* both falls held together as one rope) and hoisted. When the boat is nearly up, the falls are separated and hoisted as necessary to square the boat. When the boat is high enough, a stopper is made fast to the falls, or the lifelines are passed round the hooks and davit heads, and the falls walked back until the boat's weight is taken by the stopper or lifelines. The falls are then made fast to the bollards or cleats, and the stoppers or lifelines cast off. Wire falls fitted to special winches are usually hoisted electrically. The winches are provided with handles so that they can be also turned by hand. Before hoisting by electrical power *make sure that these handles are unshipped.*

Abandoning Ship with Passengers — Boats are lowered to the embarkation deck (usually the lowest open deck) and bowsed into the rail there with tackles at the bow and stern. Passengers are mustered in the public rooms and shepherded into the boats as soon as they are ready. They should be placed, as far as possible, in the bottom of the boat out of the way of the rowers. Side ladders are also provided to enable people to embark if necessary after boats are in the water, but it should be remembered that, in a seaway, it is most important to get the boat away from the ship's side as soon as possible to avoid damaging her against the steel plating, and they should therefore be embarked at the embarkation deck when practicable and the boat then lowered, unhooked and shoved off without delay.

LAUNCHING SHIP'S LIFEBOATS

The three types of davit shown so far in this book are all obsolete or obsolescent. They will still be found in ships for some years but new tonnage will have escape equipment that does not rely so heavily on manpower. Some of the old-style davits in ships are not sufficiently strong to support the weight of the boat and its passengers, a red band denotes Launching Crew Only.

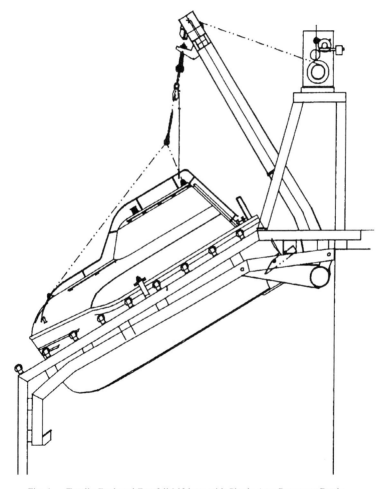

Fig. 4 — Totally Enclosed Freefall Lifeboat with Single Arm Recovery Davit.

The present Regulations require that the boat must be turned out to the embarkation deck either by gravity or by stored power. The older gravity davits were not built to conform to the requirement that the boat must go out against an adverse list of twenty degrees.

One method of complying with the rules in cargo ships is to fit a **Free Fall Lifeboat** at the stern. (Fig. 4). A davit allows the boat to be recovered if weather conditions are suitable, it can also be used for practice in port.

Deadlights — Plates fitted over portholes to strengthen them or prevent lights inside the ship from showing outboard.

Deadeye — A block made without sheaves.

Deadweight — The number of tons of cargo, fuel and stores a ship can carry.

Deeptank — A lower hold fitted with a watertight lid, which can be used for carrying ballast.

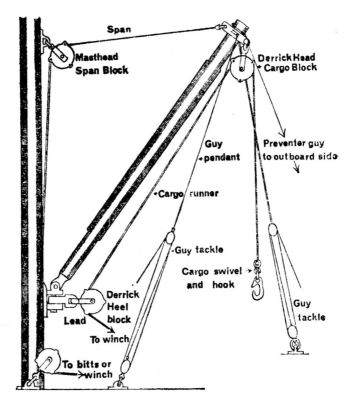

Derricks — Are used for loading and discharging cargo or stores. The heels of the derricks are usually stepped upon a "table" at the foot of the mast, 3 to 7 feet above the deck. The heels hinge and swivel on a "gooseneck". The derricks are raised by a topping-lift, or span, which consists either of a single wire shackled to the head of the derrick and leading through a block at the lowermast head, it being then shackled to a topping-lift tackle which leads down on deck, or else of a wire tackle rove between the head of the derrick and the lowermast head. They are steadied sideways by guys, consisting of short wire pendants shackled to luff tackles. Wire cargo runners or whips are led through a gin block at the head of the derrick, through a heel block at the foot of the derrick, and so to the winch.

When working cargo it is usual not to swing the derrick at each hoist, but to guy one derrick over the side and another to plumb the hatch, their runners being shackled together to one hook, called a union hook, and the cargo sling hoisted with one winch and lowered by the other. The derrick plumbing the hatch is distinguished by calling it the "stay" and the overside one the "yard" (from the old practice of sailing ships where one tackle hung on the lowermast stay and the other on the lower yardarm).

Specially strong derricks are fitted to deal with heavy lifts, stepping on the deck at the foot of the mast and being fitted with three or fourfold tackles for topping-lift and purchase.

The weight that a derrick can handle is marked on it near the heel. Some heavy derricks can lift more than a hundred tons.

DF = Direction Finder — An instrument for finding the bearing of a wireless station.

Dinghy — A small beamy square-sterned boat.

Discharge Book — A book owned by every member of the Merchant Navy showing the names of every ship served in, etc. They are handed in to the shipping master when signing-on and returned, filled in, when signing-off at the end of the voyage.

Displacement — The number of tons of water a ship displaces; in other words, her weight.

Dodger — A canvas weather screen.

Dowel — A wooden plug in a deck plank covering a bolt head.

Draw — Sails are said to draw when they are filled with wind and steady.

Draught — The depth of the keel below the waterline. This is shown by figures on the stem and stern-post which are 6 inches high, the *bottom* of the figure marking the foot.

Dunnage — Wood, mats, etc., used to secure or protect cargo from dunnage.

E
"E's for the Ensign which flies at the peak";

Ease to the Stopper — After hauling a rope sufficiently a stopper is put on temporarily. The rope is then eased slowly until the stopper has got the weight when the rope is quickly made fast to the bitts, cleat or belaying pin and the stopper cast off. (*See* **Stopper**).

E.P.I.R.B. — Emergency Position Indicating Radio Beacon.

Equator — An imaginary line running round the centre of the earth halfway between the North and South Poles, sometimes referred to as The Line, as in "Crossing the Line".

Expansion Joint — A joint across the top decks of a large passenger ship to allow for the working of that part of the ship in a seaway.

F

"F's for the Fo'c'sle, where the sailormen sleep";

Fathom — 6 feet.

Feather the Oar — To turn the blade of the oar horizontally.

Fenders — Made of cork, rope, etc., and fitted with a lanyard to defend or protect a ship's or boat's side from the quay, etc.

Fid — A wooden spike used in splicing.

Fiddle — A rail round a table top.

Fidley — The deck house over the boiler tops upon which the funnel rests.

FIRE AND FIRE-FIGHTING
Fire precautions

(1) *Cleanliness* — One common cause of fire aboard vessels is an accumulation of oil or gas in the bilges, and grease near galley stoves. Anything spilt should be cleaned up at once. Cigarette ends should be carefully stubbed out.

(2) *Petrol* is, of course, inflammable, but it is the gas (especially dangerous because invisible) freely given off by the liquid which, when mixed with the right quantity of air, becomes explosive. (This is why explosions occur in "empty" tankers, some gas from a previous cargo remaining undetected in a confined space).

(3) *Flash Point* — is the lowest temperature at which a liquid will give off vapour which can be ignited by a flame or spark. The flash point is thus a guide to the inflammability of the liquid: the lower the flash point the greater the fire hazard. When the flash point is below the local air temperature, vapour from an exposed liquid will always be present. Examples: petrol from 50° to 10°F; paraffin 100° to 120°F; diesel oil more than 130°F. Petrol is thus at all times a serious fire hazard.

Before refuelling a small boat with petrol:

(1) Stop engines.

(2) Switch off all electrical appliances, *e.g.*, fans.

(3) Extinguish cigarettes and pipes.

(4) Close all hatches, doors and potholes.

(5) Check contents of the fuel tank to avoid over filling.

(6) Close stop valve between fuel tank and engine if gravity fed.

At night, if local lighting is needed, only electric or hand lamps of safe type should be used.

While refuelling:

The delivery nozzle should be held in contact with the tank filler inlet during the whole operation. A tank should not be completely filled; some space must be left for expansion. If it is being overfilled the nozzle should be immediately shut off and held outboard.

After refuelling:

The filler cap should be tightly secured. Anything spilt should be wiped up completely, the rag or cloth being disposed of or placed where evaporation of the gas is unlikely to be a hazard. The whole boat should be opened up and allowed to ventilate for five minutes before starting any engine or lighting any stoves. Electric fans should not be used for this purpose unless they are known to be safe.

(3) *Engine Fire Hazards* — are leakage from fuel pipes, flooding or back-fires from carburettors, sparks from short circuits, overloading of electrical accessories and overheating of exhaust pipes.

When the vessel is to be left, the petrol feed should be shut off at the tank and the engine run until the carburettor is exhausted.

A drip tray, longer and wider than the engine and gearbox, should be fitted unless there is an oil-tight bulkhead fitted fore and aft of the engine.

Exhaust pipes should be water-jacketed and properly lagged.

(4) *Electrical Fire Hazards* — arise from defects in wiring or equipment, which may produce arcing, short circuits or overload. These may ignite any inflammable vapour present, or the contents or structure of the vessel

Batteries of lead-acid or nickle-alkali when being charged give off explosive gases. An accidental shorting of the terminals or the metal containers of some nickle-alkali type batteries may cause a local explosion.

(5) *Butane, Calor, and Similar Gases* — are stored under pressure and, should a leak occur in the system the gas, being heavier than air, pours out and falls to the bottom of the compartment where it mixes with the air and forms an explosive gas. Containers therefore are best kept above deck or in a separate locker so placed that any escaping gas cannot reach enclosed spaces.

When cooking is completed turn off the burner used and then *turn off the container valve* as well. The container valve should always be kept closed when the vessel is unattended.

Check for leaks at regular intervals. Liquid detergent or soapy water should be brushed round unions, valves, etc., and a watch kept for any bubbles.

When fitting a new bottle or container fit a new washer in the union nut and check that it makes a gas-tight connection. Keep some spare washers for this purpose. The outlet on the container valve has a left-hand thread, so the union nut has to be tightened in an anti-clockwise direction.

Stoves should be securley fastened down for use. Burners having a wet priming system should be located within a tray with raised rim to confine anything spilt.

Procedure for a Suspected Gas Hazard — If it is suspected that there is an accumulation of butane gas or petrol vapour present anywhere on board:

(1) Stop engines and extinguish all sources of ignition: cigarettes, pipes, cookers, heaters and lights other than electric lights.

(2) Shut the stop valves in the petrol feed and the main valve on the butane container.

(3) Open up the space affected to allow through ventilation.

(4) Try to remove the gas from the space by using a hand-operated bilge pump.

If possible:

(5) Anchor or make fast well clear of other craft.

(6) Evacuate the people on board.

(7) Get in touch with the fire brigade and, if the boat is hired, the agent or owner.

If the fire breaks out and a butane container is likely to be involved, shut the valve, disconnect the container and either remove it to a safe place or drop it overboard.

It is officially recommended that a printed notice on durable material should be permanently and prominently displayed in every boat. The suggested wording, which sums up the foregoing, is as follows:

CAUTION

Before Refuelling — *Stop engine. No smoking or naked lights. Check level in tank.*

If Petrol is Spilled — *Clean up. Ventilate boat for 5 minutes before starting engine.*

BUTANE SYSTEM

Ensure adequate ventilation when appliances are in use.

After Use — *and when the boat is unattended* — *shut off gas at container valve.*

Changing Container — *Shut off before disconnecting. Test for the leaks only with soapy water.*

SUSPECTED PRESENCE OF PETROL VAPOUR OR BUTANE

Stop engine. Extinguish all naked flames, cookers, heaters, etc., pipes and cigarettes. Shut valve on petrol feed and butane container. Ventilate entire boat.

Where possible — *Call Fire Brigade* — *or expert advice.*

FIRE

Use fire extinguishers and water from over the side.
If butane container is likely to be involved shut off — disconnect and if necessary drop overboard.
Where possible — Call Fire Brigade and assistance.

FIRE-FIGHTING

For the purpose of fire-fighting fires are classified, because different types of fires require different modes of attack.

Class A fires: paper, wood, cloth, general, rubbish.

Class B fires: inflammable liquids, petrol, oil etc.

Class C fires: live electrical hazards, motors, switches etc.

A fire can be extinguished by (1) cooling the burning material, (2) smothering or stifling the fire by excluding air from it, without which it cannot burn, or, of course, by combining these two methods.

(1)   *Cooling* — Water is an efficient and plentiful agent. A bucket of water or a 2-gallon extinguisher will extinguish a small fire of Class A. A jet of water striking a Class B fire may disperse the burning liquid and spread the fire. In this case water may be applied by a spray nozzle. This is an excellent cooling device, the heat of the fire being expended in evaporating the water while, if the compartment can be closed to prevent the entry of fresh supplies of air, smothering also occurs, the steam generated preventing air from reaching the burning liquid. Dual-purpose nozzles which can produce either a spray or a jet can be obtained.

(2) *Smothering* — (i) *Foam* — A successful means of smothering fires is the application of foam. This is produced by a mixture of water and chemicals which produce bubbles which are strong and persistent. The foam adheres to any kind of surface and being much lighter than oil will float on top of it, covering the fire and cutting off the combustible material from the atmosphere. All that is then required for the complete extinction of the fire is the cooling-off process. When possible, foam should be directed on to a vertical surface above the fire from where it will flow down and spread over the burning material. Another method of attack is a sweeping and encircling action to reduce the area of the flames, sweeping the fire into a corner before finally extinguishing it.

(ii)   *Powder* contained in an extinguisher is an effective smothering agent. Powdered chemical is forced out in a dense cloud and is particularly useful for dealing with small fires in awkward places. The powder is largely deposited on the burning material excluding the air from it while the particles remaining in the air inhibit combustion. The powder used today can deal with all classes of fire.

(iii) *Carbon Dioxide* (CO_2) is an efficient smotherer, a concentration of 30% in air being sufficient to stop combustion. It is stored under pressure and released in an invisible jet. It does not damage electrical gear or machinery, but cannot be breathed. It is most useful in electrical or galley fires of fat or oil.

(iv) *Bromochlorodifluoromethane* (BCF) excludes the oxygen and therefore smothers the fire. Stored as a liquid under pressure it causes no further damage to equipment on fire when released as a gas.

(v) *Sand*, applied by a scoop from bin or bucket, if properly used, is suitable for dealing with small fires from oil fuel drips, etc.

Large fires have small beginnings, and it is most important that all fires should be tackled *immediately they break out* while they are still small enough to be easily extinguished. For this reason, fire extinguishers, small but efficient, conveniently placed and instantly available are of the greatest value.

PORTABLE FIRE EXTINGUISHERS

Different type of extinguishers have been developed to combat the different types of fire. There are four main extinguishing agents: water; foam; dry chemical; gas. Each is expelled from the extinguisher, either by gas stored under pressure or pressure from chemical reaction, in a jet to a distance of between 5 and 50 feet in about one minute.

Water Type — The simplest type holds two gallons of water which is expelled by gas under pressure.

Foam Type — The foam is ejected by the generation of carbon dioxide gas formed by the mixing of chemicals which take place when the extinguisher is discharged.

Powder Type — Usually sodium bicarbonate, blown out by carbon dioxide contained in a cartridge under pressure.

Gas Type — Carbon dioxide, which is heavier than air, sinks in it and will penetrate inaccessible places, but cannot be breathed, is stowed under pressure in a liquid state. On release the liquid gasifies at the nozzle expanding to some 250 times the volume of the liquid.

BCF — which is a by product of the petro-chemical industry, has the advantage that it is stable when heated in a fire, unlike the old carbon tetrachloride which could in certain circumstances produce phosgene gas. All smothering gasses are dangerous especially in confined spaces.

In order to make it easy to recognise the contents of each extinguisher, there is a colour-code which many countries use.

RED — denotes WATER
CREAM — denotes FOAM
BLUE — denotes POWDER
BLACK — denotes CO_2
GREEN — denotes BCF.

REGULATIONS

The kind of firefighting equipment carried varies with the type of the ship. Merchant ships are governed by the Merchant Shipping (Fire

Appliances) Regulations 1980. For the purposes of these regulations, ships are divided into twelve categories from One which is for Foreign going Passenger ships to XII which is for some of the larger Pleasure craft.

Pleasure yachts have guidance for safety issued by the Royal Yachting Association. Pleasure yachts used commercially, particularly those used for chartering, have Codes of Practice drawn up after consultation with the trade and issued by the Department of Transport. These codes cover stability, watertight integrity and lifesaving appliances as well as fire precautions for small commercial craft.

A typical medium sized cargo ship of Class VII will have a Fire pump in the engine room and an emergency fire-pump outside the engine room. Either of these pumps will supply water to hoses connected to a fire main. The hydrants on this main must be so arranged that two jets of water may be brought to bear on a fire in any part of the ship including the engine room and shaft tunnel. Some of the hose nozzles will be dual purpose, jet or spray and all new hoses must be made of rot proof material although some of the old flax canvas hoses may still be in use. In case there is a fire in port and the local fire brigade is called, there must be an international connection to allow the shore party to use the fire main. There will be suitable portable fire extinguishers in the accommodation and working spaces. Oil fired ships will have a large fixed foam extinguisher to cover any area to which the burning oil might run. The engine room will also have its own extinguishing system probably CO_2 flooding the space after it has been evacuated.

There will be a number of Firemen's outfits each of which will be complete as follows:—

FIREMEN'S OUTFITS
Ships of 500 tons or over are required to carry firemen's outfits. These consist of:
  (a) a breathing apparatus,
  (b) a portable self-contained electric battery-operated safety lamp which will give a light for at least three hours,
  (c) a fireman's axe.

Breathing Apparatus — May be either:
  (a) A smoke helmet or a smoke mask, connected by an air hose to an air pump or bellows which is worked by someone still on the open deck in clean air. Smoke helmet gear consists of the helmet, air tube, lifeline, and bellows. To rig it, connect one end of the air-tube to the nozzle of the bellows, and the other end to the nozzle on the tubes, at the back of the helmet. Open the windows in front of the helmet and then put it on. In the top of the helmet is fitted a small valve for the escape of excess and used air. Always see that the small black disc of this valve is in place. The soft leather

bottom of the helmet should be worn inside the jacket. The waistbelt is fitted with an arrangement at the back for holding the air-tube in position and to prevent it from dragging at the helmet. Make the lifeline fast. Before going into foul air, work the bellows to see that the air is coming into the helmet all right and then close the windows and valve (if fitted) in front of the helmet. Work the bellows at about 25-30 strokes per minute.

The signals used by the man wearing the smoke helmet are:

1 tug on the lifeline = "I want more air".

2 tugs on the lifeline = "Slack away lifeline".

3 tugs on the lifeline = "Help me out".

(b) A self-contained breathing apparatus comprising one or more cylinders of compressed air providing at least 24 cubic feet of free air. The pressure is automatically reduced and regulated so that the air may be easily breathed. A pressure gauge shows the air pressure in each cylinder and an audible warning is given when 80% of the air has been consumed. Spare cylinders are kept fully charged.

With the breathing apparatus is:

(a) A safety-belt and harness, to which is connected a snap-hook.

(b) A fire-proof life-and-signalling line, at least 10 feet longer than the hose length, of copper or galvanised steel wire rope and overlaid with a covering surface to provide a surface which can be firmly grasped.

(c) A helmet and protection for the eyes and face against smoke.

(d) Protective clothing of material capable of protecting the skin from the heat radiating from the fire and from burns and scalding by steam.

(e) Boots and gloves of electrically non-conducting material.

(f) Non-inflammable plates bearing the code of signals, one attached to the safety harness and the other to the free end of the lifeline.

LARGE FIRE-FIGHTING SYSTEMS

In ships, additional to a number of portable extinguishers, fire pumps and hoses, more complete systems of fire protection may be fitted.

For open spaces, e.g., cabins and large rooms — the Sprinkler system is used: for compartments which may be inaccessible because occupied by possibly combustible material, e.g., cargo holds and store rooms — a Smothering Gas is employed.

Sprinkler System — Small-bore pipes are led overhead throughout the accommodation, running from a water tank where the water is kept under pressure. At short intervals along the pipes sprinkler heads are fitted. Each sprinkler head is fitted with a plug valve kept in place by a small bulb containing liquid. When the heat increases the liquid expands and when a certain temperature has been reached the liquid bursts the bulb and the plug valve is released. Water pours out of the pipe and is broken into a heavy spray which scatters over a wide area.

An alarm system is started when a sprinkler operates and the water in the supply tank is kept topped up by a pump (started automatically when the water has fallen to a certain level) supplying sea water. The system is initially filled with fresh water to prevent corrosion. The whole sprinkler system is thus completely automatic and is very prompt in action.

Carbon Dioxide Smothering System — Gas smothering systems are usually combined with fire detection systems. Small-bore pipes are led from the compartments to be served *e.g.*, cargo holds — to a detection cabinet, generally on the bridge, and thence to an extractor fan. Air is continuously sucked from the holds and passes through the detector cabinet. Should a fire occur smoke will also pass through the cabinet and will be detected, either by eye or a light-sensitive cell which rings an alarm when smoke intercepts the light falling upon it.

The valve attached to the pipe from where the smoke is coming is then closed. This action first closes the pipe to the detector and then opens the supply line to the gas cylinder room. Gas can then be supplied by opening the gas cylinders in turn.

All ventilation and ventilating fans to the fire area must of course be stopped, and, provided that the gas can be retained in the space, it is a very effective extinguisher.

It is vitally important that warning be given to anyone in the space about to be flooded with gas, and they should leave directly it is heard. It is possible to connect the alarm so that it must be sounded before the gas can be turned on.

Inert Gas Generators — In ships with large compartments to protect; for instance tankers whose empty tanks may still contain some inflammable gas, a large quantity of gas may be required. Fuel oil is burnt in a combustion chamber in the right quantity to produce an exhaust gas exceedingly low in oxygen. After cooling, the exhaust gas is fed to a distribution manifold and used in the same way as the CO_2 gas supplied in cylinders. The gas can be produced for as long as fuel and water are obtainable *See* **The Merchant Shipping (Fire Appliances) (Amendment) Regulations 1993.**

Fire Plan — A plan showing all the fire fighting equipment in the ship must be readily available to any shore based fireman.

Fish — To strengthen a spar by lashing another to it.

FLAGS

National Flags — A ship wears, or flies, the colours of her country at the stern, either from a flagstaff or the peak of a gaff. Ketches and yawls, if without gaffs, fly it from the mizzenmast head. Our national flag is the Union Flag (or Union Jack as it is commonly called) which is made up of

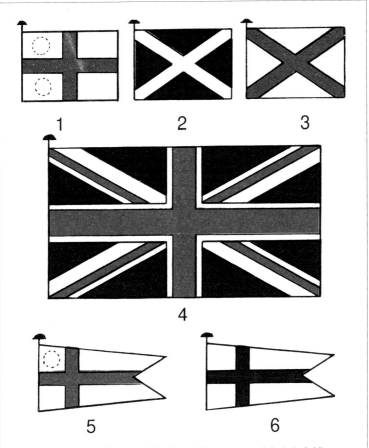

1. Cross of St. George of England. Flown by an Admiral. A Vice-Admiral flies this flag with one red ball in the upper canton. A Rear-Admiral has a ball in each canton.
2. Cross of St. Andrew of Scotland.
3. Cross of St. Patrick of Ireland.
4. Union flag comprised of the crosses of the 3 patron saints. Flown at the main by an Admiral of the Fleet and worn at the Jackstaff forward by all H.M.S. when at anchor or when dressed.
5. Commodore's Broad Pennant. Commodores, 2nd class, have a red ball in the upper canton.
6. Broad Pennant flown by a Commodore of a Convoy.

(1) the Cross of St. George of England (a red cross on a white ground); (2) the Cross of St. Andrew of Scotland (a white diagonal cross on a blue ground); (3) the Cross of St. Patrick of Ireland (a red diagonal cross on a

white ground). The broad part of the white diagonal should be on top of the red next to the mast.

The Red, White and Blue Ensigns were originally used to denote the divisions of a fleet, but today the White Ensign is worn by the Royal Navy and the Royal Yacht Squadron; the Blue Ensign by the Royal Naval Reserve and certain Yacht clubs; and the Red Ensign by merchant ships and other yachts.

Most of the Dominions' and Colonies ships wear the Blue or Red Ensigns with their badge on the fly. For example: Australia bears the four stars of the Southern Cross and two other stars; New Zealand the four stars of the Southern Cross alone; Canada, however, has a flag vertically divided, red, white and red, with a red maple leaf on the white. South Africa has a horizontal tricolour striped orange, white and bright blue, with three small flags (the union Jack and the flags of the Transvaal and Orange Free State) in the white. Merchant ships are sometimes registered in countries other than that to which they belong and they then fly a "Flag of Convenience"; the flag of the country of registration. Examples: Panama; Liberia (a red and white striped flag with a single white star in a blue canton).

When at anchor ships may also fly a jack at the jackstaff in the bows. Ships of the Royal Navy wear the Union Flag, and merchant ships the Pilot-jack (a Union Jack with a white border round it).

Ships may fly other flags in addition to their national ensign:

The Royal Standard — When the Queen is aboard a ship she flies at the mainmast-head the Royal Standard, the four quarters of which bear the Lions of England and Wales, the Lion of Scotland, and the Harp of Ireland.

Admirals' Flags — An Admiral of the Fleet flies the Union Flag at the main-mast head. An Admiral flies the flag of St. George; a Vice-Admiral, and Admiral's flag with a red ball in the top corner next to the mast; and a Rear-Admiral, an Admiral's flag with a red ball in each of the quarters next to the mast. A Commodore flies a Broad pennant (a broad white swallow-tailed pennant with a red St. George's cross thereon).

Other ships of the Royal Navy fly at the main mast-head a narrow white pennant with a red St. George's cross.

House-Flags — Merchant ships fly at the main masthead the flag of the company to which they belong. They sometimes also hoist a small house flag at the jackstaff instead of the Pilot-jack.

Yachts fly the burgee of their club.

In port, ships often fly at the foremast-head, as a compliment, the flag of the country in which they are, or to which they are about to sail. This is known as a Courtesy Flag.

The Royal Mail pennant (white, with a crown over a bugle between the words "Royal Mail" in red) is flown by ships having Her Majesty's mails on board.

EXAMPLES OF HOUSE FLAGS AND CLUB BURGEES

BRITISH STEEL plc

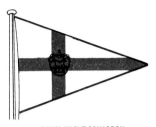

ROYAL YACHT SQUADRON

P. & O. LINE

ROYAL CRUISING CLUB

B.T. (MARINE) LTD

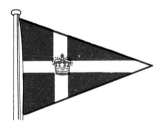

ROYAL THAMES YACHT CLUB

STAR OFFSHORE SERVICES LTD

ROYAL CLYDE YACHT CLUB

Racing Flag — A square flag flown in place of a yacht club burgee when a yacht is engaged in a race. Other vessels, as an act of courtesy, endeavour to keep clear of yachts racing.

On special occasions ships are "dressed" rainbow-fashion. Flags are stopped to long dressing-lines which are then hoisted to the mastheads.

Flags — Flags can be "broken" from the masthead. The flag is tightly rolled up and the tack is passed round it and the bight hitched. The flag is then hoisted and the halyards made fast. A jerk on the tack halyard will then break the flag free. Be careful to hoist the toggle of the flag close up to the truck; a flag improperly hoisted 6 inches, below the truck or with the fly foul of the halyards, gives the whole ship a slovenly appearance.

Colours are hoisted in port at either 0800 or 0900 until sunset; but when entering or leaving port at any time it is light enough for them to be seen, and at sea when meeting men-o'-war or ships belonging to the same line. In the last two cases a merchant ship salutes by dipping her ensign (lowering her ensign until the other ensign is dipped in reply, and then hoisting it again). The Red Ensign is always dipped to a White Ensign, and when two merchant ships of the same company meet the junior captain's ship dips to the senior. Ensigns are flown at half-mast as a sign of mourning.

Flag halyards must be slacked in rainy weather as they shrink when wet. (*See* also Signals and Signalling).

Flake — To coil a rope in layers so that it will run clear.

Flare — The overhanging of the bow plating.

Flatten in the Sheets — To haul aft the sheets and so flatten the sails.

Fleet — To draw the blocks of a tackle apart ready for another pull on the same rope. To change the position of anything.

Flemish Coil — To coil a rope down flat so that each flake lies exactly outside the next.

Flush Deck — A deck which goes without a break from stem to stern.

Fore-and-aft — Lengthwise of a vessel.

Fore-and-aft Rig — A vessel that is not square-rigged, *i.e.*, which has no yards.

Forecastle — The raised part of a ship's hull forward; the crew's quarters. (Pronounced foke-sel).

Forecastlehead — The deck over a forecastle.

Forward (pronounced for'ard) — Towards the bow.

Foul Hawse — Two cables crossed when both anchors are down.

Frap — To pass a rope round anything to draw or keep it together.

Free — Sailing with the wind; not close-hauled.

Freeboard — The height of the ship's upper deck above the waterline.

Full and Bye — Sailing close to the wind, but keeping the sails full.

G

"G is for the Galley, where the cook hops around",

Gale — A wind blowing at more than 33 knots.

Gale Warnings — Consist of a black canvas cone which looks like a triangle when hoisted. The South Cone (point downwards) is hoisted for gales commencing from a southerly point; such gales often veer sometimes as far as N.W. For gales coming from the E. or W. the South Cone will be hoisted if the gale is expected to change to a southerly direction. The North Cone (point up) is hoisted for gales beginning from a northerly point or for gales which are expected to change to a northerly point. With the scaling down of the Coastguard these old signals are unlikely to be shown.

Gangway — A narrow way aboard a ship. Hence gangway-ladder, usually known as the gangway.

Galley — The ship's kitchen.

Gantline — A rope rove through a block at the masthead used for hoisting anything aloft.

G.M.D.S.S. — Global Maritime Distress and Safety System.

G.P.S. — Global Positioning System. Satellite Navigation. (American).

GLONAS — Global Navigation Satellite System. Satellite Navigation. (Russian).

Grapnel — An anchor with more than two flukes and no stock.

Greenwich Observatory — The place from which all longitude is measured and times regulated.

Gripes — The lines which secure a boat to the deck or a griping spar.

Guest Warp — A rope led from forward to the bottom of the accommodation ladder or gangway to help boats coming alongside.

H

"H for the Halyards we haul up and down".

Halfdeck — The cadets' or apprentices' quarters in a merchant ship.

Handling Craft — **General** — Vessels must always obey the Rule of the Road. A good look-out must always be kept and the proper lights or signals shown. In narrow channels power-driven vessels to keep to the starboard side of the fairway and, if manoeuvring to avoid approaching craft, make appropriate sound signals. (Page 135).

Each craft has her own peculiarities, due to her construction, loading and trim, and all are affected by the tidal stream or current, wind, and the transverse thrust of the propeller as well as the action of the rudder. These points will now be considered.

Draft — A deeply laden vessel will be slower to answer her rudder and slower to lose way, but she will be less affected by the wind. When she is light she is quicker to respond to the rudder (provided the rudder is still deeply enough immersed), but more affected by the wind and therefore may be difficult to handle.

In very shallow water a vessel "smells the ground" when she may not answer her rudder. She tends to swing first towards the shallow side and then away.

Trim — Most vessels are trimmed slightly by the stern, the rudder and propeller being then well submerged, but if too much by the stern they will be difficult to turn into the wind.

Tide — or more properly, the tidal stream, moves the ship bodily over the ground and the slower the ship's speed through the water the greater the effect of the tide upon her. It is thus essential to know in which direction the tide is flowing. This can be found by referring to a Tidal Atlas or to the chart; it can be seen when passing buoys (which appear to be moving against the tide) and by noting which way anchored ships are pointing (providing that they are not wind rode).

Before anchoring or berthing the ship alongside she should stem the tide (unless the wind is stronger than the tide), and a good allowance should be made for the tide's movement if the ship has to be turned round.

The Wind — The effect of the wind on a power-driven ship must not be neglected at *slow* speeds. When engines are stopped and the vessel loses way her bow will probably fall away from the wind; when she has lost all way she may lie with the wind a little abaft the beam. When she is moving astern the stern tends to swing into the wind: "the stern seeks the wind's eye".

58

The Transverse Thrust of the Propeller — The propeller exerts a turning effect upon the ship. This effect is particularly noticeable with large-diameter propellers; where the ship is light; and at slow speeds.

A merchant ship in ballast may have part of the propeller blades out of the water at the top of the stroke, when, of course, their effect on the water will be nil. The propeller then acts like a paddle wheel. The paddle-wheel effect is still present even when the top of the blades are submerged. A right-handed propeller tends to cast the stern to starboard when the engine is going ahead because the lower blade meets with more resistance than the upper blade.

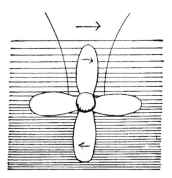

Fig. 1 — **Right-handed screw going ahead** — The lower blade acting like a paddlewheel, tends to cant the stern to starboard and the bow to port.

If when viewed from astern, a propeller turns right-handed, as do the hands of a watch, when the engine is turning ahead, the propeller is said to be right-handed. Single-screw ships are usually fitted with a right-handed propeller. Twin-screw ships usually have a right-handed starboard propeller and a left-handed port propeller.

The stream of water passing the rudder makes the rudder effective and when the ship is going ahead at her usual speed the rudder easily overcomes the tendency of the screw to cant the stern sideways. When, however, the ship is stopped or nearly stopped, the turning effect of the propeller becomes greater than that of the rudder. This effect can be made use of when turning the ship short round.

The Action of the Rudder — is greatest when the angle it makes with the fore-and-aft line is about 35° and the propeller is turning full speed ahead, and will be zero when the propeller is stopped and the ship is not moving through the water. When the ship is moving astern the effect is much less certain, though if the rudder is put to port the stern should swing to port, and if put to starboard should swing to starboard.

A ship when moving ahead pivots round a point about one-third of her length from forward; about where the bridge is usually placed, and if moving astern, about one-third of her length from the stern. The ship thus skids round and the skidding action can be increased and the ship be made to turn more quickly by using the *engine* at full speed when the *ship* is moving slowly.

Turning the Ship Round — The turning circle (*See* Fig. 2) is more precisely a turning spiral. For a large merchant ship using full rudder and moderate speed the advance might be 500 yards and the transfer 700 yards.

Turning Short Round — In a river or harbour where there is not room to turn round without going alternately ahead and and astern with the engines, the transverse thrust of the propeller is used to maximum effect. An anchor dropped under foot and the cable then braked will also help to snub the ship's head round.

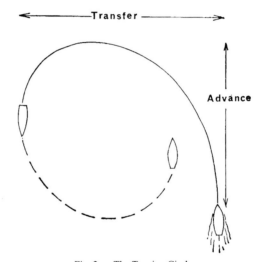

Fig. 2 — The Turning Circle.

Advance: usually about three to four ship's lengths.
Transfer: usually slightly greater than the advance.
Path traced by stern: the ship will move at least two ship's lengths before the stern clears the original path.
Time to turn through 90° about 2 or 3 minutes. Loss of speed in turning 90° or more, about half.
Loaded ships will have a larger turning circle and take longer to turn than light ships.
Yachtsmen should know the manoeuvring capabilities of large ships so that they can foresee their future actions. Yachts turn more quickly of course.

If a turn is made to port the screw helps to turn the ship as long as the engine is going ahead, but directly it is put astern the propeller tends to cant the stern back to port and eventually overcomes the effect of the rudder. If the turn is made to starboard the rudder overcomes the tendency of the propeller to turn the ship to port while the engine is going ahead and directly the engine is put astern the action of the lower blade helps the ship to continue to turn. Single-screw ships must therefore be turned round to starboard. The only exception to this is when there is much wind on the starboard beam. When ships are moving astern their sterns tend to fly into the wind and therefore with a fresh wind on the starboard beam the wheel is put to *port* and when the ship has started to swing the engine is put astern (and the wheel amidships). The stern still swings to windward and, when pointing into the wind, the wheel is put to port again and the engine ahead.

With twin-screw ships the turn is best made stern to wind. If turning to port, the port engine is put astern and the starboard engine ahead, the transverse thrust of the propellers assisting the turning movement.

Stopping a Ship — The distance that a ship will advance from the time of putting her engine astern from (full ahead) until she is stationary is between 6 and 12 ship's lengths, depending upon the initial speed, displacement and engine power. The time taken will probably be between 4 and 10 minutes.

The rudder will lose its effect as the speed of the ship decreases and the bows will either fall off to starboard or, if the wind is strong, away from the wind.

Stopping in an Emergency in Shallow Water — One anchor, or much better, both anchors let go underfoot and the brakes then screwed up will check the way of the ship. The windlass is strongly secured and, with the cables leading aft, much of the stress comes on the hawse pipes; but stand clear of the cables in case one parts.

High Speed Craft — Rough water can be dangerous at high speed, for the faster you go the harder you hit the waves. Good judgement is required in determining how fast a large and powerful ship may be driven into a heavy sea without risk of sustaining damage. Smaller high-speed craft are very responsive to the throttle and lose speed quickly; a vigilant look-out can detect the approach of exceptionally high seas (for it is not the succession of waves which is likely to cause damage, but rather the odd one or two which are larger than, or out of phase with, the others) and speed be reduced to meet them.

A good look-out ahead is essential at all times. Flotsam is difficult to see, especially in rough weather and at night, and a high-speed collision with driftwood may be disastrous.

High-speed craft are fitted with small rudders to reduce drag which, at low-speed, may be rather ineffective. Nonetheless, by varying the speed of the port and starboard propellers these craft are very manoeuvrable.

Approaching an Anchorage or Berth — If the harbour is strange to you the "Pilot" and chart should be studied carefully before arriving, conspicuous objects and depths of water noted, the direction of the wind and tide (and which of the two is likely to have the stronger influence on the ship when she has slowed down). You should decide what you want to do and how you intend to do it, but be prepared to alter your plan if some circumstance; for example, a ship anchored in your way, compels you to do so.

Reduce speed in good time for two reasons:

(1) If it is necessary to stop while still going full ahead the ship will travel for four to nine ship's lengths (the bigger the ship the longer the distance) before coming to rest and during that time will be partly out of control, her head probably slewing to the starboard. If she is going slow the full astern action of the propeller will stop her much more quickly.

(2) If the ship is moving slowly ahead and it is desired to turn quickly to port or starboard the wheel can be put over and the engines put, temporarily, to full speed ahead. The propeller will push water rapidly past the rudder and the ship will turn quickly without gaining any appreciable speed through the water, the skidding action being increased. It has been said "Full speed is fool speed".*

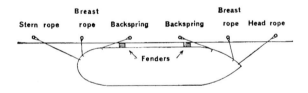

Fig. 3 — Mooring Lines.

Speed through the water can be judged by looking over the side; speed over the ground by noting the rate at which objects ashore are being passed.

Approach the berth up tide or, if the wind is stronger, up wind. Heading into the tide the ship will have steerage way even when almost stationary with relation to the ground and be perfectly under control.

An anchor should always be ready for letting go however it is intended to secure the ship.

---

* This and other quotations, are from R. A. B. Ardley's valuable book, *Harbour Pilotage*.

Securing Alongside — See that heaving lines, hawsers and fenders are ready for use. Approach the berth at an angle of between 10° and 30°, depending upon the direction and strength of wind. It is easier to berth port side to the quay because the engine when put astern cants the ship's head away from the berth. If berthing starboard side to the quay the angle of approach must be small because the bow will swing in when the engine is put astern. With a fresh on-shore wind the off-shore anchor may be dropped abreast of the berth to prevent the ship's head falling too quickly to leeward and to assist in hauling off the berth when leaving.

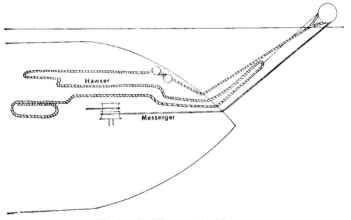

Fig. 4 — Doubling up a Head Rope.

The headrope is usually the first rope sent ashore, immediately followed by the back-spring, to prevent the ship going too far ahead and then the stern lines. It is, of course, of great importance to keep the stern lines clear of the propeller. Once the ship is in position alongside, the moorings can be strengthened either by adding other ropes or by passing out the bights of the ropes already used with the aid of a messenger, so that the mooring is trebled; a process known as "doubling up". (*See* Fig. 4).

Remember that if the berth is in a tidal dock or river the moorings may have to be slackened or hove in as the level of the water changes.

Leaving a Berth — First, single-up the moorings (take inboard any bights or extra hawsers, leaving the single ends still fast ashore). See that the engines and engineers are prepared. The manoeuvre will depend upon the direction of the tide and wind, the up-tide end of the ship being got out first. If there is no tide get the stern out first. (*See* Figs. 5 and 6).

(1) *No Tide* — Heave the forward backspring taut; let go aft; turn the rudder towards the shore; put the engine to slow ahead. The stern will swing out, and when far enough, come astern on the engine; let go forward and back away.

(2) *With the Tide Aft* — Single up to a backspring forward and a breast rope aft. When the breast rope is slackened the tide will force the stern out, the backspring preventing the ship from forging ahead. The engine is then put astern.

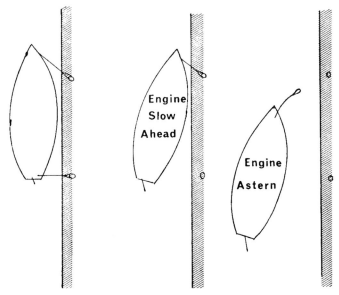

Fig. 5 — Leaving a Berth with no tide: Stages 1st, 2nd and 3rd.

(3) *With the Tide Forward* — The bow must come out first. Single up to an after backspring and forward breast rope. When the breastrope is slackened the tide will force the bow out. The bow must not be allowed to come too far out or the stern of the ship will come on to the quay; the swing can be checked by holding on to the breastrope. The engine is put ahead and the ship steamed out.

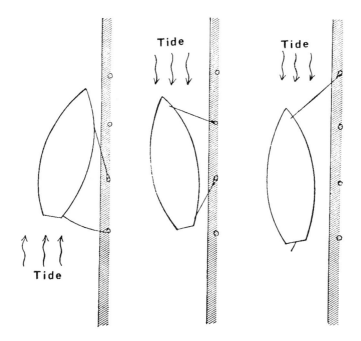

Fig. 6 — Leaving a Berth with a tide.

If the quay is so low that the ship's quarter will ride over it, the propeller will be endangered if the stern comes on to the quay. In this case, single up as before but with a head rope in addition. Let go breast rope and backspring, turn the rudder away from the shore and go slow ahead. The head rope will prevent the stern from swinging in.

BAD WEATHER

In Harbour — (1) *Alongside* — See that the moorings are sufficient, bearing an equal strain and well parcelled, and that the fenders between ship and quay are properly placed.

(2) *At Anchor* — If the anchorage is exposed, and particularly if the wind is on shore (*i.e.*, blowing towards the shore), it may be better to weigh anchor and get to sea and gain as much sea room as possible. If staying in the anchorage, let go a second anchor as follows: Put the rudder hard over to sheer the ship away from her anchor and when she has gone as far as she will go drop the second anchor and then pay out both cables to the greatest possible scope. Spreading the anchors in this way helps to stop the violent sheering about that causes the anchors to drag or the cables to part.

At Sea — In very bad weather a large power-driven vessel is hove-to by slowing the engines and putting the rudder over, so that their combined action is to keep the vessel just moving through the water and shouldering the sea with her port or starboard bow (port bow is best, as the transverse thrust of the screw will help to push her bow into the wind), in which position she will lie most safely.

Smaller vessel must decide upon the best action to take. If there is an easily accessible harbour to leeward, run for it. If this is not possible, it is better to be in open water away from the influence of the land.

If one cannot run for shelter there are four possible plans:

    (1) To heave-to.

    (2) To lie a-hull.

    (3) To lie to a sea-anchor.

    (4) To run with warps streamed aft.

(1) *Heaving-to* — If it is important not to drift down wind because, for example, there is land to leeward, the vessel may be hove-to. The aim is to keep the sea on one bow or the other, just making headway but not too much leeway, so that the vessel moves slowly about 90° from the wind. Having shortened sail the sheets are trimmed with the mainsail flattened in and the foresail probably sheeted to windward, and the helm lashed a little to leeward.

(2) *Lying a-hull* — This is the simplest method. All sail is furled, the helm lashed, and the vessel left to take up her own position in the sea, giving rather than standing up to them. The drift to leeward will be considerable; and the motion violent, the vessel rolling heavily for she will be more or less beam on to the sea. Oil can be of great assistance in preventing waves from breaking on board, though it does not of course reduce their size.

(3) *Lying to a Sea Anchor* — Some vessels will lie to a sea anchor more easily than others. A modern yacht with a short keel and single mast may not do so; a ketch or a yawl should lie much easier with a scrap of canvas hoisted aft. Drift to leeward is reduced, the vessel lies nearly head to sea with a relatively easy motion.

(4) *Running with Warps Streamed Astern* — Plenty of sea room is required for this method. The warps steady the vessel, keeping her stern to the sea and preventing a broach-to.

Some ships, trimmed by the stern, run well.

If there is risk of drifting on to a lee shore, let go an anchor (lower it, if the water is deep) and pay out all the cable. "Never go ashore with an anchor at the bow". Even if the anchor does not hold it will bring the ship's head into the wind and cause her to drift less rapidly to leeward. With plenty of room to leeward, but in shallow water, the cable paid out without the anchor, will act as a sea anchor and help to keep the ships bows to the sea. In a boat a sea anchor can be used for this purpose.

MAN OVERBOARD

If the ship is moving slowly into a strong head wind and sea it is probably quicker to stop her by reversing the engines. At all other times it is better to turn round. Procedure in the latter case: Throw a lifebuoy. Put the wheel hard over (if the man is seen to fall from *e.g.* the starboard side, put the wheel to starboard to swing the stern and propeller away from him). Reduce speed. Detail a look-out to do nothing else but keep the man in sight and keep pointing at him. Man the lee boat and lower it when the ship has nearly completed a full turn and lost almost all headway, provided that you can still see the man or lifebuoy. (*See* Fig. 2 Turning Circle). Page 60.

A small vessel can be manoeuvred to go alongside the man. Have a heaving-line and boathook ready, and approach at slow speed with propeller stopped.

If sailing, and beating or reaching, the helm is put up and the vessel wears round, approaching close-hauled on the opposite tack and luffing up alongside him.

If running free, stand-on briefly, then down helm, go about and tack back to luff up alongside him.

Hand the Log — To haul the log inboard.

Hands — The crew of a vessel. 'One hand for the ship and one for yourself'. Hang on! Good advice in bad weather.

Handsomely — Slowly, with care.

Handspike — A bar used as a lever.

Handy Billy — A small handy tackle.

Hanks — Metal rings for keeping the luff of a sail to a mast or stay.

Harness Cask — A cask for holding meat.

Hatchway — An opening in the deck.

Hatch — The cover of a hatchway.

Hawser — A strong rope used for mooring a ship, etc.

Heave-to — To stop a vessel by bringing her head into the wind.

Heel — To lean over.

Helm — The tiller.

H.M.S. — Her Majesty's Ship.

Hogged — A vessel that is strained so that her ends have dropped is said to be hogged.

Holiday — A space left when painting.

Holystone — A stone used to scrub the deck.

Horse — An iron bar athwart the stern for the main-sheet to travel on.

Hull Down — When a ship's hull is hidden by the horizon and only her upper works are visible she is said to be "hull down".

I

"I's for the Irons where the stunsail booms ship".

Inert Gas — A system where inert (non-explosive) gas is used to fill a space which might otherwise be filled with explosive petroleum gases. See page 52.

Irish Pennant — A rope yarn or rope's end hanging untidily from aloft.

J

"J's for the Jibs right ahead of the ship".

Jacob's Ladder — A rope-ladder with wooden rungs. Used in conjunction with a painting stage.

Jettison — To throw overboard cargo to lighten a vessel.

Jumper Stay — The wire stay running between foremast and funnel.

Junk — Old condemned rope. (2) A Chinese or Japanese sailing vessel.

Jury Rig — A temporary rig in place of a rig carried away.

K

"K is for Keelson and also for Keel".

Killick — A small anchor.

King Plank — The centre-line plank of a planked deck.

Knife — A seaman should never be without a knife. Sheath knives are carried at the back of the belt.

Knot — A speed of one sea mile per hour.

L

"L is for Log with its line and its reel".

Landfall — A ship is said to make a landfall when she first sights land after an ocean passage.

Lanyard — A short line used for securing anything.

Latitude — The distance that a ship is north or south of the equator.

Launch — To drag an object along; not only a boat into the water.

Leads and Lead Lines — When a vessel is near the coast it is very important that she should know how much water is beneath her keel. The process of finding the depth of water is known as **sounding**, and in shallow water of less than 20 fathoms (a fathom is 6 feet) the hand lead and lead line is used unless an echo sounder is fitted.

Hand Lead and Line — The lead for a ship usually weighs 14 lbs and is fitted with a strop or becket: the line is 25 fathoms long with an eye-splice in one end. For a yacht a 10 fathom line is usually long enough, with a 7 lbs lead. The line is bent to the lead by passing the eye through the strop and then the lead through the eye.

When marking a lead line the line must first be wetted and stretched. A plank of the deck is marked at 2, 3 and 5 fathoms by which to measure the line.

Markings — When marked in fathoms the lead line is marked in the following way:

At 2 fathoms a piece of leather with 2 tails
At 3 fathoms a piece of leather with 3 tails
At 5 fathoms a piece of white calico
At 7 fathoms a piece of red bunting
At 10 fathoms a piece of leather with a hole in it
At 13 fathoms a piece of blue cloth
At 15 fathoms a piece of white rag
At 17 fathoms a piece of red bunting
At 20 fathoms a piece of cord with 2 knots

If a longer line than 25 fathoms is used, 30 fathoms is marked by 3 knots, 40 by 4 knots, etc., and a single knot put at each 5 fathoms. The fathoms that are not marked are known as "deeps".

The soundings on new Admiralty charts (and on all continental charts) are in metres, instead of fathoms, and when these are used, it is convenient to mark the lead line in metres also.

The markings used by the Hydrographic Service, which follow the older fathom markings, are as follows (except that blue and white bunting is used at 8 and 18 metres instead of yellow). The two 2-colour marks can be thought of as coming one metre before 5 and 10 metre marks.

Markings in Metres

   1 and 11 metres — one strip of leather
   2 and 12 metres — two strips of leather
   3 and 13 metres — blue bunting
   4 and 14 metres — green and white bunting
   5 and 15 metres — white bunting
   6 and 16 metres — green bunting
   7 and 17 metres — red bunting
   8 and 18 metres — yellow bunting
   9 and 19 metres — red and white bunting
   10 metres   — leather with a hole in it
   20 metres   — leather with a hole in it and 2 strips of leather.

The following jingle may be useful in memorising the sequence of the colours:

   3 is blue and
   5 is white,
   6 as green as a starboard light;
   7 as red as a nose that's cold,
   Pieces-of-**8** are yellow as gold.

Heaving the Lead — In a ship the lead is hove from a small platform on each side of the ship called the chains. Before going into the chains see that (1) the apron against which you will lean is properly secured; (2) the lead line is clear and its end made fast. Lean well out into the apron and grasping the lead line between the second and third fathom mark, swing the lead backwards and forwards like a pendulum to gain momentum, and when the speed of the ship requires it, swing the lead two full turns over your head. Having thus got up speed the lead is hove forward into the water. As the ship moves forward the lead line leads straight up and down and the leadsman calls out the depth, the fathom being named last, *e.g.,* "By the mark seven", "And a half seven" ($7\frac{1}{2}$ fathoms), "A quarter less eight" ($7\frac{3}{4}$ fathoms), "Deep eight", or whatever it may be.

When anchoring, the lead is dropped on the bottom to show when the ship stops and gathers sternway, the leadsman singing out "Ship stopped" or "Ship going astern".

In a Sailing Yacht — Stand on the starboard side of the foredeck by the shrouds. Make the end of the lead line fast to a shroud or stanchion and coil the line loosely on deck. Secure yourself so that you can use both hands. Take a coil equal to about twice the depth of water expected in the left hand. Hold the lead and about two fathoms of line in the right hand and lower the lead over the side until it is just clear of the water. Swing the lead like a pendulum and release it on the forward swing so that the lead flies ahead of the yacht before entering the water. Allow the line to run out over the hand and when the line is up and down and the lead on the bottom call out the sounding. Bump the lead gently to be sure that it is on the bottom.

The Echo Sounder — This is an electronic device which sends out little packets of high-frequency soundwaves and picks them up when they are reflected from the bottom. Sound travels at a known rate through the water, so the time taken for the echo to reach the vessel gives the depth of water. The depth is shown on a dial, and since a number of packets of soundwaves are sent out each second, a continuous record of the depth is indicated.

Leeboards — Boards fitted to the sides of flat-bottomed sailing craft. The lee one is lowered down when sailing close-hauled to prevent her drifting to leeward.

Lee Side — The side further away from the wind.

Leeway — The drift of a vessel to leeward when sailing near the wind.

Life Saving Appliances — There are Rules made by the Government for all British Vessels except yachts less than 45 feet long (which does *not* mean that yachts of less than 45 feet should not carry Life Saving Appliances).

The requirements for L.S.A. are laid down in The Merchant Shipping (Life-Saving Appliances) Regulations 1986 and the Amendments made in 1991. As in the Fire Fighting Regulations the ships are divided into twelve categories. The examples taken here are for a medium sized cargo ship of Class VII with a crew of say 15.

There will be a *Lifejacket* for every person on board. This will be designed to turn the wearer on his back in the water and keep his mouth clear of the water even if he becomes unconscious. It will have a whistle, a light and reflective patches. It will be clearly marked DoT.

Other lifejackets marked BS 3595 are more convenient for use in yachts but are not approved for general use in Merchant ships.

The Civil Aviation Authority also approve lifejackets but these are not generally found in ships.

Anything other than these three standards is a buoyancy aid and may be suitable for dinghy sailing but it is not a lifejacket in which to abandon ship.

There will be eight *Lifebuoys* marked DoT. Two of these will be in quick release slips on the wings of the bridge, they will be fitted with *Manoverboard Lights and Smoke*.

Two others will be fitted with *Lights* and two will be fitted with *Lines*.

The ship will have a *Totally Enclosed Lifeboat* under davits on each side of the ship, each one to carry all the persons on board.

Instead of the two lifeboats mentioned the ship may have one *Totally Enclosed Freefall Lifeboat* fitted at the stern. This boat to carry all the persons on board.

A TOTALLY ENCLOSED BOAT.

In addition, if two boats are fitted there must be *Inflatable Liferaft* capacity for all the persons on board on each side of the ship. If only the freefall lifeboat is fitted, the liferaft on at least one side of the ship must be fitted with a launching appliance.

Unless a lifeboat is fitted as a *Rescue Boat*, a rescue boat must be carried. In a small ship this might be a *Rigid Inflatable Boat*. A free-fall boat is not considered suitable as a rescue boat owing to the possible difficulty of recovering it in bad weather.

Every *Lifeboat* must be fitted with a diesel engine such that the fully loaded boat can motor at 6 knots for 24 hours. It must also be capable of towing a liferaft loaded with 25 persons at 2 knots.

Lifeboat falls must be fitted with simultaneous disengaging gear operated from the coxswain's position. In a cargo ship it must be possible for the full complement of persons to be able to board the boat in three minutes.

One Survival craft *EPIRB* must be available on each side of the ship.

Three two-way *Radio Telephone* sets must be provided and a *Portable Radio station* unless a fixed *Radiotelegraph* is installed in a boat.

The boat-deck and the water overside must be adequately lit by lights powered by the emergency generator.

Lifeboat Equipment

Sufficient buoyant oars to make headway in a calm sea, crutches for these oars.

Two boathooks.

A buoyant bailer and two buckets.

A survival manual.

A binnacle with an efficient compass which is either illuminated or luminous.

A sea anchor.

Two painters at least 15 metres long, one attached to the quick release gear forward.

Two hatchets one at each end.

Three litres of water per person the boat is certified to carry (or 2 litres and de-salting apparatus).

Three rustproof drinking vessels, one graduated in millilitres

Food rations totalling not less than 10,000 kilojoules for each person the boat is certified to carry.

Four rocket parachute flares.

Six hand flares.

Two buoyant smoke signals.

One waterproof electric torch with spare batteries and a spare bulb.

One daylight signalling mirror with instructions for use.

One copy, Department of Transport Rescue Signal Table.

One whistle.

A First Aid outfit.

Six doses of anti-seasickness medicine and one sick-bag per person the boat is certified to carry.

A jack-knife on a lanyard.

Three tin openers.

Two buoyant rescue quoits with 30 metres of buoyant line.

A manual bilge pump.

One set of fishing tackle.

Sufficient tools for minor adjustments to the engine.

Two portable fire extinguishers suitable for oil fires.

A searchlight.

A radar reflector.

Thermal protective aids (survival bags) for 10% of the number of persons the boat is certified to carry.

Passenger Ships. Because they are subdivided into much smaller compartments than cargo ships and because they can carry out counter flooding to correct a list, ships of Class I carry lifeboats for the number of persons on board, 50% on either side of the ship and liferafts for 25% of the persons on board. This basic requirement is varied according to the trade in which the ships are engaged.

Ferries may have *Marine Escape Systems* in addition to a proportion of the lifeboats. Fast ferries may only have liferafts and a rescue boat or boats. Some small passenger ships in local waters, required to have liferafts, may substitute *Open Reversible Liferafts* which do not have canopies and may accommodate as many as sixty-five persons. The LSA Regulations for Passenger Ships of Classes III to VI(A) were made in 1992.

Liferafts; Buoyant Apparatus; Lifeboats.

Liferafts — Inflatable liferafts are inflated by a gas which is not injurious to people and takes place automatically, usually when the line connecting the liferaft to the ship is pulled. A topping-up bellows or pump may be provided. The buoyancy is arranged in a number of separate compartments, half of which when inflated can support the number of persons which the liferaft is fit to carry, so that there is a reasonable margin of buoyancy if the raft is damaged or partially fails to inflate. The construction includes a cover of a highly visible colour which is automatically set in place when the raft is inflated. The cover protects the occupants from exposure and also provides a means of collecting rain. The top of the cover is fitted with a lamp which is activated by salt water and a similar lamp is fitted inside the liferaft. At each opening means is provided to enable persons in the water to climb on board. A lifeline is fitted both inside and outside the liferaft and it has a painter and can be towed.

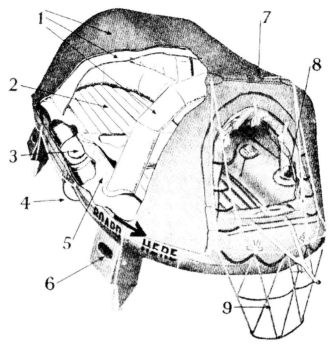

Fig. 1 — INFLATABLE LIFERAFT. Normally stowed on board in a cylindrical valise. Launched by being thrown overboard (this can be done by 2 or 3 men), or the raft will float free from a sinking ship. Inflation is automatic. Made in various sizes to carry up to 25 persons.

1. The double-walled canopy provides complete protection from wind, sea and sun. It is automatically erected by the inflation of two arches.

2. Inflatable floor insulates occupants from sea in cold regions.

3. Bow and stern sections of the main buoyancy chamber are sealed from each other by baffles. If one half becomes badly damaged the other will support the full complement of men.

4. High pressure tube from CO_2 gas inflation cylinder stowed under raft.

5. The main buoyancy chamber is inflated at a pressure of only 2 lbs per sq. in. Deflation is therefore slow in the event of a leak which can easily be repaired. The single large diameter tube buoyancy chamber provides great rigidity.

6. Water pockets which can be retracted for towing, give added stability.

7. Light unit shows position from air or sea.

8. Non-return valve seals arch from main buoyancy chamber after inflation.

9. Boarding net and hand ladders to permit easy entrance from the water.

Liferafts are stable in a seaway and they can be easily righted by one person if they should inflate upside down. (*See* Fig. 1).

An inflatable liferaft should be properly surveyed every 12 months at an approved service station.

Rigid liferafts only differ from the inflatable type in that they are usable and stable when floating either way up. They occupy more space, of course, than inflatable liferafts when stowed on board, but do not require frequent surveys.

Equipment

   (a) Two sea anchors, one permanently attached to the liferaft and one spare with line.

   (b) For liferafts which are fit to accommodate not more than twelve persons; one bailer, two sponges and one safety-knife.
       For liferafts which are fit to accommodate thirteen persons or more; two bailers, two sponges, and two safety-knives.

   (c) One topping-up pump or bellows.

   (d) One repair kit capable of repairing punctures in the buoyancy compartments.

   (e) One rescue quoit attached to at least 30 metres of line.

   (f) Two paddles.

   (g) Two parachute distress signals complying with the provisions of Part I of the Eighth Schedule to the Merchant Shipping (Life-Saving Appliances) Rules.

   (h) Six hand flares complying with the provisions of Part II of the Eighth Schedule to the Merchant Shipping (Life-Saving Appliances) Rules.

   (i) One waterproof electric torch suitable for Morse signalling, together with one spare set of batteries and one spare bulb in a waterproof container.

   (j) One daylight signalling mirror and one signalling whistle.

   (k) Food rations totalling not less than 10,000 kilojoules for each person the raft is permitted to carry.

   (l) Watertight receptacles containing $1\frac{1}{2}$ litres of fresh water for each person whom the liferaft is fit to accommodate, of which $\frac{1}{2}$ litre per person may be replaced by a suitable de-salting apparatus capable of producing an equal amount of fresh water.

   (m) One rustproof drinking vessel, graduated in $\frac{1}{2}$, 1 and 2 oz.

   (n) Three safety-tin openers.

   (o) One fishing line and six hooks.

   (p) Six seasickness tablets for each person whom the liferaft is fit to accommodate.

   (q) A first aid outfit complying with the provisions of Part II of the Seventh Schedule to the Merchant Shipping (Life-Saving Appliances) Rules.

   (r) Instructions printed in the English language on how to survive in the liferaft, and Rescue Signal Table.

   (s) Thermal protective aids (survival bags) for 10% of the persons the raft is certified to accommodate.

Launching — Liferafts may be thrown overboard, when they will inflate, the gas being released by a sharp pull on the painter. The painter should be kept made fast to the vessel. They can be inflated first, the passengers then embarked, and then lowered into the water. This may be done as follows:

Launching Liferafts by Davit — The crew consists of one man in general charge and four others whose primary duties are:

No 1 — In general charge.

No 2 — Davit operator.

No 3 — Tending right-hand bowsing line.

No 4 — Tending left-hand bowsing line.

No 5 — Leading Hand of Liferaft.

"Abandon Ship" — The liferaft crew close up at the launching station.

"Prepare to Launch" — No. 2 swings out the davit arm and releases the hook to the suspended position.

Nos. 3 and 4 remove section of guard rail.

No. 5 opens liferaft stowage.

Nos. 3, 4 and 5 carry the first liferaft from stowage to launching position, making sure that the bowsing line pockets face inboard.

Nos. 3 and 4 hook on, 3 holding the liferaft shackle and 4 closing the hook.

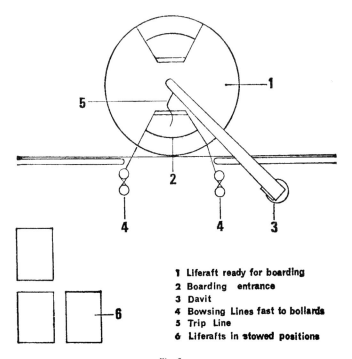

1 Liferaft ready for boarding
2 Boarding entrance
3 Davit
4 Bowsing Lines fast to bollards
5 Trip Line
6 Liferafts in stowed positions

Fig. 2.

No. 5 makes sure that the hook safety catch is properly engaged and that the trip line is free and lying inboard across the valise.

Nos. 3 and 4 remove the bowsing lines from the pockets, pull taut and make them fast to the bollards.

No. 1 checks that these operations have been carried out.

"Heave Away" — No. 2 heaves away to pre-set mark on the davit fall. The raft automatically inflates and swings out.

Nos. 3 and 4 loosen the bowsing lines at the bollards and keep in hand. When the liferaft is completely inflated they bowse it close alongside and make fast the bowsing lines.

"Leading Hand Aboard" — No. 5 boards the liferaft, checks that both buoyancy chambers and the central tube are firm and then reports: "ready for boarding".

"Passengers Aboard" — No. 1 ensures an orderly approach.

No. 3 and 4 assist passengers to board.

No. 5 distributes the passengers, the first to the outboard entrance and the remainder alternately to the left and right.

"Cast Off" — Nos. 2 and 3 cast off the bowsing lines and throw them into the boarding entrance. *Both lines must be inside.*

No. 5 holds the trip line from the hook safety catch firmly.

No. 2 stands by to lower.

"Trip" — No. 5 pulls trip line, listening for the "click" which tells him that the hook will release the liferaft as soon as it is waterborne.

"Lower" — No. 2 releases the brake on the davit. The liferaft will lower at a pre-set rate and be automatically unhooked an reaching the water.*

Buoyant Apparatus is a term meaning flotation equipment other than liferafts and buoyant jackets, designed to support persons who are *in* the water. Passenger ships are, in some cases, permitted to carry Buoyant Apparatus in place of some lifeboats or liferafts.

Buoyant apparatus is usually made in the form of seats for passengers, stowed on the upper deck, from which they would float off should the ship founder and from which they may be easily thrown overboard. They are constructed to withstand being dropped into the water (as are liferafts) and to be stable whichever way up they float. Buoyancy is provided by buoyant material (*e.g.,* polyurethane foam) or air cases. Grablines with cork or wood floats are fitted all round the apparatus by which persons in the water can support themselves. A painter is also fitted.

Liferafts are an effective means of saving life and give good protection against the elements, but they are not navigable. This is often unimportant because, should a ship have to be abandoned it will be usually only after an SOS message has been broadcast and received. The best thing for the survivors to do in this case is to remain in the position signalled in the SOS

---

* The above drill is included by permission of its originator, Commander N.F. Keene, D.S.C. RD. R.N.R.

message until they are rescued by a vessel coming in answer to their call for assistance.

Lifeboats are designed to carry the people on board the parent ship if she founders; to rescue people from another vessel in distress; and to rescue a person who has fallen overboard. They are therefore built to be very stable, seaworthy commodious and navigable. They have internal buoyancy in the form of either air cases or buoyant material and this must be at least 10% of the cubic capacity of the boat and sufficient to float her and her equipment when the lifeboat is flooded and open to the sea, so that the top of the gunwale amidships is not submerged.

The dimensions of the boat are cut in the gunwale and the number of persons which the boat may carry is also shown.

All modern lifeboats are propelled by diesel engines. There are many types of boat to cover the needs of the various trades and dangers arising from them.

Partially enclosed and Self-righting partially enclosed boats are found in Passenger ships and may hold up to 150 persons. Totally enclosed boats are found in cargo ships. Totally enclosed boats with water spray fire protection are found in tankers. Totally enclosed boats with air support systems are found in chemical tankers. All ships have at least one designated rescue boat with special equipment for the boat and crew.

LIFE SAVING EQUIPMENT FOR YACHTS 13·7 METRES IN LENGTH OR OVER

(Laid down in the Merchant Shipping (Life-Saving Appliances) Rules 1986.

Yachts of 13·7 metres to 21·3 metres in length, inclusive.

(a) **Liferafts** — One or more of sufficient aggregate capacity to accommodate the total number of persons on board.

(b) **Lifebuoys** — At least two.

(c) **Lifejackets** — One for each person on board.

Notes: Liferafts shall be so stowed that they can be readily transferred to the water on either side of the ship.

If the yacht does not proceed to sea, or which only proceeds to sea during the months of April to October, inclusive, on voyages in the course of which it is not more than 3 miles from the coast of the United Kingdom, she need not carry a liferaft but shall carry *lifebuoys* at least equal to half the total number of persons on board. The minimum number of lifebuoys to be carried is two. Any such ship that operates only in smooth waters shall not be required to carry more than two lifebuoys.

A Buoyant Line — At least 10 fathoms in length.

Distress Signals — Six parachute or six red star distress signals.

Yachts of 21·3 metres in length or over shall carry

(a) **Liferafts** — At last two, of sufficient aggregate capacity to accommodate *twice* the total number of persons on board.

(b) **Lifebuoys** — At least four.

(c) **Lifejackets** — One for each person on board.

(d) **A Line-throwing Apparatus** — (*See* "Line-throwing Apparatus").

(e) **Distress Signals** — Six parachute or six red star distress signals.

For yachts of 24·3 metres in length and over

In addition: a rescue boat or an inflated boat served by a launching appliance.

SAFETY EQUIPMENT AT SEA — CHECK LISTS

These lists have been drawn up by the Royal Yachting Association and are reproduced with their permission.

The following are suggested as the minimum requirements for yachts and motor cruisers:

(a) **Personal Equipment** — Buoyancy and (*i.e.* lifejacket) with whistle and/or light attached, for each person on board. Children and non-swimmers should always wear lifejackets when afloat.

A knife and spike, on a lanyard, for each member of the crew.

For any yacht proceeding out of sheltered waters: a lifejacket (to British Standard 3595) and a safety-harness (to B.S. 4224) for each of the crew.

(b) **For Day Cruising and Occasional Nights at Sea** — Horseshoe lifebuoy with self-igniting light; fire extinguisher(s); Flares, say six red and six white; First-Aid Box; Radio receiver capable of obtaining weather forecasts; Radar reflector; Mariner's Compass, correctly adjusted; Anchor, with chain and warp; fixed Bilge pump, with a bucket and bailer to supplement it; Leadline or Echo-sounder; two torches with spare bulbs and batteries; Chart of the area, and local tide tables; a reliable Watch or Clock.

(c) **For Off-Shore Cruising** — As for (a) and (b) above, plus six hand flares, red; six hand flares, white; six parachute rockets, red; a second horse-shoe type lifebuoy, with self-igniting light and a flag on a stave; additional First-Aid equipment; additional anchor or kedge; self-inflating liferaft with emergency pack (or inflatable dinghy for smaller craft); and additional and independent bilge pump; Radio receiver suitable for shipping frequencies, Coastal Radio Stations and Radio Direction Finding; Gas detector (if petrol or calor gas are stowed between decks); Charts for the areas concerned; Nautical Almanac and Tide Tables; (Deck Watch and Sextant if astral navigation is to be used); List of Lights; Tidal Atlas, Hand bearing Compass; Patent log (or electrical equivalent); International code flags; Aldis light, or similar; Barometer.

They add as the minimum requirements for a yacht racing as a cruising yacht:

Strong and effective single or double **lifelines or guardrails** of adequate height fitted along the sides of the yacht from bow to stern. A dinghy stowed in a seamanlike manner on deck. This may be of rigid or collapsible type and should have adequate non-inflatable buoyancy. Alternatively, an **inflatable dinghy**, which shall either be self-inflating or be carried fully inflated, may be used. The dinghy shall be capable of carrying the whole crew. Alternatively, the yacht shall carry both dinghy and liferaft complying with Royal Ocean Racing Club Seaworthiness and Safety Regulations. Navigation lights and foghorn.

ADDITIONAL SAFEGUARD

Yachtsmen making a coastal passage in U.K. waters are strongly advised to contact the local H.M. Coastguard station to complete **Coastguard "66" Passage Report Cards**. This ensures that the Coastguard Service has full particulars regarding the craft, occupants, its equipment and intended voyage, for use should search and rescue action be needed. There is no charge for this service.

Lifeboatmen's Certificate — Certain regulations call for Certificated Lifeboatmen, the present day equivalent of this certificate is the Certificate of Proficiency in Survival Craft.

To qualify for a Certificate of Proficiency in Survival Craft
An applicant must:
  (a) be not less than 18 years of age;
  (b) have attended an approved basic sea survival course;
  (c) have performed not less than 12 months sea service, subject to three months remission which they may be granted in respect of completion of an approved survival craft training course;
  (d) pass an examination, conducted or approved by the Department of Transport.

EXAMINATION SYLLABUS

Candidates will be required to satisfy the Examiner:
  (a) by practical demonstration as required, their ability to carry out the tasks listed in PART I below; and

(b) that they are familiar with the matters relating to the launching and operation of survival craft and survival at sea as listed in PART II below.

PART I

1. Each candidate will be required to act as the person in charge and also take part in the practical operation of launching and boarding of survival craft, clearing the ship or dock side quickly, handling whilst afloat and disembarkation from and the recovery of survival craft.

2. Candidates will be expected to know the allocation of duties for the craft in use, the orders commonly used in the operations of launching, handling and recovery, and in particular the specific commands 'STILL' and 'CARRY ON'.

3. Each candidate will be required to demonstrate that he is able to don a lifejacket correctly.

4. Each candidate will be required to demonstrate that he can interpret the markings on a survival craft with respect to the number of persons it is permitted to carry.

5. Candidates will be required to demonstrate that they are able to row, steer, erect a mast, set the sails, manage a boat under sail and steer by compass.

6. Candidates will be required to demonstrate that they are able to use signalling equipment, including pyrotechnics, and portable radio equipment for survival craft.

PART II

1. **Emergency situations**
   1. Collision.
   2. Fire.
   3. Stranding.
   4. Foundering.

2. **Musters, drills, abandonment**
   1. Emergency and abandon ship signals.
   2. Action to be taken when signals are made.
   3. Duties assigned to crew members in muster list.
   4. The value of training and drills: the need to be ready for any emergency.

3. **Survival procedure**
   1. Actions to be taken when required to abandon ship.
   2. Actions to be taken when in the water.
   3. Actions to be taken when aboard a survival craft.
   4. Main dangers to survivors.
   5. Methods of helicopter rescue.

4. **Life-saving appliances**
   Life-saving appliances and arrangements in general use on board ships.

5. **Types of survival craft**
   The construction and outfit of the following survival craft, together with their particular characteristics and facilities.
   1. Lifeboat.
   2. Enclosed lifeboat.
   3. Class C boat.
   4. Inflatable boat.
   5. Davit-launched inflatable liferaft.
   6. Hand-launched inflatable liferaft.
   7. Rigid liferaft.
   8. Any other survival craft in general use.

6. **Operation of survival craft**
   1. Preparation.
   2. Launching, including methods of launching into a rough sea.
   3. Boarding.
   4. Clearing the ship's side.
   5. Actions to be taken after leaving the ship.
   6. Coming alongside.
   7. Disembarkation.
   8. Beaching.

7. **Types of davit and their methods of operation**
   1. Gravity.
   2. Luffing.
   3. Single Arm.
   4. Any other type of davit in general use.

8. **Operation of davits for recovery of boats.**
   1. Manual means.
   2. Compressed air.
   3. Electricity.
   4. Any other method in general use.

9. **Means of propulsion**
   1. Rowing.
   2. Sailing.
   3. Motor: methods of starting and operating survival craft motors and their accessories.
   4. Hand propelling gear.

10. **Survival craft handling**
    1. Handling in rough weather.
    2. Use of emergency boats and motor lifeboats for marshalling liferafts and rescue of survivors and persons in the sea.

11. **Instructions**
    the instructions provided with and attached to survival craft and their equipment.
12. **Use of survival craft equipment**
    1. Protective covers.
    2. First Aid Kit
    3. Painters.
    4. Sea anchors and drogues.
    5. Fire extinguishers.
    6. Radio devices including approved emergency position indicating radio beacons.
    7. All other equipment.
13. **Medical**
    1. The effect of hypothermia and its prevention including the wearing of protective garments to maximum advantage.
    2. The effects of dehydration and heat exposure.
    3. Resuscitation techniques.
    4. Dealing with injured persons during and after abandonment.
14. **Provisions**
    The apportionment of food and water carried in survival craft.

Light-to — To lighten or lift a rope towards the man who is making it fast.

Lightvessels — Used to mark important shoals. They are generally painted red and have their names painted on their hulls in large white letters.

Limbers — Holes beneath the timbers of a boat or floors of a ship to allow any water in the bottom of the vessel to drain fore-and-aft.

Line-throwing Appliances — The appliance includes four rockets and four lines minimum 4mm diameter, with a breaking strain of not less than 2 kilonewtons. The range of the rocket and line is 230 metres in calm weather. The lateral deflection on either side of the direction of firing should not exceed one tenth of the range.

In some types the rocket and line are contained in a case with a handle, sighting tube and trigger, the upper part of the case holding the rocket, the lower part the line. In other types the rocket is fired from a pistol. The line is, of course, attached to the rocket by a fire-proof trace.

List — To heel or lean over to port or starboard.

Lizard — A piece of rope fitted with a thimble on one end.

Loadline — A line 18 inches long running through a ring cut and painted on both sides of a merchant ship amidships. A ship may be safely loaded until this line is awash. A line for use in fresh water is cut a few inches above and forward of the loadline. The loadline is often called the "Plimsoll Mark", after Samuel Plimsoll who brought it into use.

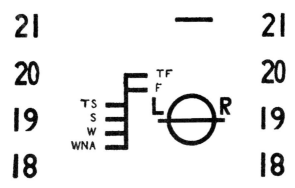

LOADLINES AND DRAUGHT MARKS ON A MERCHANT SHIP'S SIDE.

The upper line is the Deck line. TF = Tropical Fresh Water. F = Fresh Water.
TS = Tropical Summer. S = Summer. W = Winter. WNA = Winter North Atlantic.
LR = Lloyd's Register (other initials for other Assigning Authorities).

The line and circle are the original "Plimsoll Mark".

The figures on the bow and stern denote the amount of water the ship is drawing. The figures are 6 inches high and the bottom of the figure represents the foot. If marked metrically each decimeter (10cm) is shown.

Logs — The log is an instrument which shows the distance a ship travels through the water, and sometimes the speed also. Walker's Ship-Log (which is one often used) consists of a brass rotator fitted with vanes which cause it to turn a certain number of times in each mile travelled. It is towed astern by a specially made **log line** about 35-40 fathoms long. At the inboard end the log line is connected to a **governor** or flywheel, which helps the log to rotate at a uniform speed, and the governor is connected to the **register** fitted with the necessary wheel-work and a dial which shows the distance run. The face of the register usually consists of a large dial marked from 0 to 100 miles and two smaller dials, one showing hundreds of miles and the other tenths.

For yachts a very small rotator and a thin waxed terylene line is used.

The log may be towed from the taffrail, the register shipping in a fitting specially provided for it, or it may be towed from a boom projecting over the side of the ship to keep the rotator from being fouled by anything thrown overboard. Sometimes an electrical or mechanical connection is fitted to the register so that the log can be read in the chartroom or on the bridge.

Logs must be "handed" or hauled in before the ship stops. To hand the log, catch hold of the log line abaft the governor and disconnect it from the governor, then as you haul in the log line, pay it out again on the opposite quarter until the rotator is on board. Then haul in the line again, coiling it down left-handed on the rotator. This is necessary to get the turns out of

Walker's Log.

the line. Another method is to disconnect the log line from the governor and hook a swivel on to the line and run it forward along the deck, coiling it down afterwards as before. When the log is streamed from the boom amidships a grapnel is thrown over the log line near the end and the rotator is lifted on board first. The boom is then swung forward, the line disconnected and coiled down.

Great care must always be taken to prevent the rotator knocking against the ship's side for, if the vanes are bent, the log will not register accurately.

After the log is handed it should be hung up to dry before being stowed away, and the rotator wiped over with an oily rag. The register should be kept well lubricated, the cover sliding back easily to allow it to be oiled while running.

Bottom Logs — Are sometimes used instead of towing logs. An impeller is mounted on the bottom of the vessel and is rotated by the passing water so many times per mile; the speed and distance run being shown on dials. In ships, the impeller or log probe, can be withdrawn into the ship before entering harbour; in yachts it is accessible from the bilge and can be cleared from inside the vessel if it gets fouled by weed.

Log Book — A diary kept aboard all ships, written up at the end of each watch and showing courses steered, weather, etc., etc.

Longitude — The distance a ship or place is east, or west, of the Greenwich meridian, measured along the equator.

Lookouts — The duty of a lookout is to report lights or objects sighted to the officer of the watch. This is usually done by striking the bell in the following manner:

1 stroke = A light or object in sight on the starboard bow.
2 strokes = A light or object in sight on the port bow.
3 strokes = A light or object in sight right ahead.

The direction may also be reported as "right ahead" or as so many points on the starboard or port bows. Royal Navy practice is to report the bearing as "Red" or "Green" followed by the number of degrees (in tens) from ahead, so that a ship seven points on the starboard bow would be reported as "Green eight oh" (80°). The bearing is given first, then a description of the object (ship, smoke, etc.), followed by either "close" or "distant", *e.g.,* "Green three oh, a periscope, close".

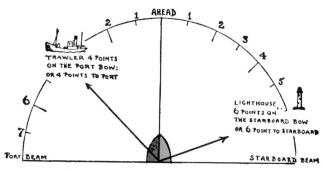

To report the Approximate Bearing of an Object
in Terms of Points on the Bow.

Lookouts should also report that the steaming lights are burning brightly every half-hour when the bell is struck.

Loom — The loom of a light is its reflection in the sky while the light itself is hidden below the horizon.

(2) Also of an oar. The opposite end to the blade.

M

"M is for the Miles we must sail wet or shine".

Manropes — Ropes forming the handrails of a gangway.

Marline — Two-stranded, left-handed tarred hemp, smaller and more tightly laid up than spunyarn.

Marine Escape System — An inflated chute leading to an inflated floating platform from which survivors may board the waiting liferafts.

Marline-spike — A steel spike used in splicing etc. When working aloft the marline-spike should always be secured round the neck with a lanyard.

Marry — To join the ends of two ropes, or to hold two ropes together which are to be hauled on at the same time.

Master — The official title of the captain of a merchant ship.

Master-at-Arms — A chief petty officer responsible for discipline.

Messenger — A light rope used for hauling a heavier one.

Midships — In the middle of the ship.

M.S. — Motor ship.

M.V. — Motor vessel.

N

"N is for Neptune, we'll meet on the Line".

Nautical or Sea Mile — 6080 feet (1853 metres). 10 sea miles — $11\frac{1}{2}$ land miles.

Neap Tide — A tide that does not rise or fall as much as the average tide.

O

"O's for the Orders that send us away".

Oakum — Old hemp rope unlaid and with the yarns picked out separately; used for caulking, etc.

Offing — To seawards; towards the horizon.

Overhand Knot — A knot made by placing the end of a rope over its standing part and then pulling the end through the bight so formed.

Overhaul — To overtake; to overhaul a purchase is to lengthen it by hauling the blocks away from each other.

P

Paint — Is used to protect as well as decorate material. More than one coat is necessary. The surface must be clean, smooth and dry and the paint well rubbed on. Primer and undercoats prevent rusting. Anti-fouling paint is used below the water-line and is applied over the anti-corrosive paint. One gallon covers about 320 feet and lasts for about 10 months.

Palm — A leather guard and thimble worn on the hand when sewing canvas.

Panting — The tendency of the bow plating to move in and out when driving into a heavy head sea.

Paravane — An instrument for clearing mines from the path of a ship. One paravane is towed on each bow of the ship, the point of tow being right forward and the paravanes towing out on each bow at an angle of about 30° from the fore-and-aft line.

Parbuckle — A method of hoisting a spar or cask when no derrick, etc., is available. A strop having been made fast, the bight of the strop is passed under the quarters of the cask and the end brought back onboard. The end being hauled on, the cask is rolled up in the bight.

Peggy — A boy, or ordinary seaman, told off to act as steward to a seaman's mess.

Pendant — A length of wire rope to which a tackle is usually shackled.

Pennant — A long, thin, tapering flag also spelt pendant.

Podger — A short steel bar pointed at both ends used as a lever, etc.

Poop — The raised after part of a ship.

Port — The left side of the ship looking forward. The port side-light is red. Boats, cabins, etc., on the port side are always numbered with even numbers. (*Aid to memory*. Port — red — left — are all shorter words than starboard — right — green).

Porthole — A circular window in the ship's side or in a bulkhead. Known as a scuttle in the Royal Navy.

Pratique — Licence to a ship to trade with a place, granted on the ship being declared to have a clean bill of health.

Pulpit — The guardrail round the bow of a yacht. If there is one round the stern it is known as a Pushpit.

Purchase — A tackle, *e.g.,* a 3-fold purchase.

Q
"Q's Quarantine at the end of the trip".

Quadrant — A flat plate in the form of a quarter of a circle, or quadrant, attached to the head of the rudder stock to which the steering chains or steering engine is connected.

Quarantine — Regulations concerning the health of any persons arriving in any country by sea or air.

Quarter — A ship's sides near the stern. On the quarter; in a direction between right aft and abeam.

Quarterdeck — The after end of the upper deck. One should always salute when stepping on to the quarterdeck of a man-of-war.

Quartermaster — The chief duties of a quartermaster are, at sea, to steer the ship (taking two-hour tricks at the wheel), and in port, to tend the gangway.

R

"R is for Rum; we'll take a good nip".

Rake — The leaning of masts towards the bow or stern.

Rank — The rank of officers is shown by stripes of gold lace worn on the sleeves or on shoulder straps. In the Royal Navy uniform, the top stripe is formed into a curl above the stripes. The Royal Naval Reserve lace is the same as that for the Royal Navy, but contains the letter "R" within the curl. Sea Cadet Corps and Combined Cadet Force officers wear Wavy stripes.

Distinction Lace.

1. Royal Navy — Captain.
2. Royal Navy Reserve — Lieutenant-Commander.
3. Merchant Navy — Chief Officer.
4. Sea Cadet Corps — Sub-Lieutenant.

The Merchant Navy lace consists of stripes with a diamond formed in the centre of the stripes. The branch or department to which officers belong is shown in some cases by coloured cloth worn between the

95

Rank 96

stripes: *Royal Navy*—medical officers, red; constructors, grey; *Merchant Navy* — engineers wear purple; surgeons, red; pursers, white; radio officers, green. Officers of and above the rank of commander and masters of merchant ships wear gold leaves on their cap peaks.

Royal Navy

Rank	Number of Stripes
Admiral of the Fleet	4 and 1 broad one
Admiral	3 and 1 broad one
Vice Admiral	2 and 1 broad one
Rear Admiral	1 and 1 broad one
Commodore	1 broad one
Captain	4
Commander	3
Lieutenant-Commander	2 and 1 narrow one between them
Lieutenant	2
Sub-Lieutenant	1
Midshipman	White patch on coat collar
Naval Cadet	White twist on coat collar

Merchant Navy

Rank	Number of Stripes
Master (Captain) and Chief Engineer Officer	4
Chief Officer and Second Engineer Officer	3
First Officer	2 and 1 narrow one
Second Officer and Third Engineer Officer	2
Third Officer and Fourth Engineer Officer	1

Ratings — *Royal Navy* — Fleet chief petty officers: badge; Royal Coat-of-Arms, Chief petty officers: specialist badge on lapel of coat and three buttons on cuff. Petty officers: crown and crossed anchors. Leading rates: anchor. These, and Good Conduct and specialist badges are worn on the right arm. Stars and crowns are added to these Branch Badges to show the degree of qualification. (*See* pages 98, 99 and 100). In the *Merchant Navy* boatswains wear on the left arm crossed anchors; boatswains' mates, a single anchor; carpenters, crossed axes; carpenters' mates, a single axe; quartermasters, a steering wheel; engine-room leading hands, a propeller.

CAP BADGES

1—OFFICERS,
ROYAL NAVY.

2—PETTY OFFICERS,
ROYAL NAVY.

3—OFFICERS,
MERCHANT NAVY.

4—PETTY OFFICERS,
MERCHANT NAVY.

BADGES WORN BY RATINGS IN THE ROYAL NAVY

──────── BERET BADGES ────────

| Warrant Officer | Chief Petty Officer | Petty Officer | Junior Ratings |

──────── OPERATIONS BRANCH ────────

| Diver | Mine Warfare | Radar | Sonar |

| Radio Operator (General) | Radio Operator (Tactical) | Tactical Systems (Submarines) | Electronic Warfare |

| Seaman | Survey Recorder | Missile | Weapons Analyst |

BADGES WORN BY RATINGS IN THE ROYAL NAVY

SPECIALIST BADGES

Marksman

Airborne Missile
Aimer

Seacat Aimer

Aircraft Controller

Commando

Navigator's
Yeoman

Subsunk Parachute
Assistance Group

Submariner

Aircrew

Cabin Attendant

Parachutist

Seaman assigned to duties
on the Royal Yacht Britannia

BADGES WORN BY RATINGS IN THE ROYAL NAVY

GOOD CONDUCT BADGES

4 Years

8 Years

12 Years

— COXSWAIN (SM) — BRANCH

C.P.O Coxswain   P.O. and below

—— MEDICAL BRANCH ——

(Letters Under)
N - State Registered Male Nurse
R - Radiographer
P - Physiotherapist
H - Health Inspector
L - Laboratory
M - Mental Nurse
HP - Health Physicist
PD - Pharmacy Dispenser

— P.T — BRANCH

Physical Training

—— ENGINEERING BRANCH ——

(Letter Under)
M - Mechanical
L - Electrical

Marine Engineering Mechanic

(Letter Under)
O - Ordnance
R - Radio

Weapon Engineering Mechanic

The new Warfare Branch Badge is the same as the current Weapons Engineering Badge with additional identifying letters. This Branch is now being introduced and will subsume several other branches over the next few years.

—————— RATE BADGES ——————

Warrant Officer   Chief Petty Officer   Petty Officer   Leading Rate
(Buttons on Cuff)

Rattle Down the Rigging, To — To put ratlines (rope steps) upon the shrouds. The end of each ratline is seized to the forward and after shrouds and the ratline is clove-hitched round the shrouds between.

Reeve — To pass the end of a rope through a block, etc.

Render — A rope is said to render when it slips round the bitts to which it has been secured.

Rescue Boat — All seagoing ships must now have one or two rescue boats. These may be designed for the purpose or may be lifeboats designated as rescue boats.

The crew must have suitable lifejackets and immersion suits.

Ride — To lie at anchor or secured to a buoy.

Riding Light — An anchor light.

Rocket Life-Saving Apparatus — If a vessel is wrecked in a place where rescue by the Rocket Life-Saving Apparatus is possible, a rocket with a thin line attached will be fired across the vessel. Get hold of this line and when you have made the end fast, signal to the shore.

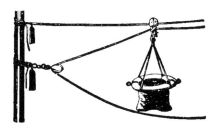

Using the Hawser and Whip with the Breeches Buoy.

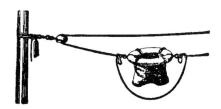

Using the Endless Whip without the Hawser Breeches Buoy.

When signalled to do so, haul on the rocket line until you get a tail block with an endless fall rove through it.

Make the tail of the block fast to a place well above the deck, so that the line will not chafe on anything, and leaving a place above the tail-block for the hawser (*see* next paragraph) to be made fast to. Unbend the rocket line from the whip. When the tail-block is fast and the rocket line unbent from the whip, signal to the shore as before.

When this signal is seen, a hawser will be bent to the whip and hauled off to the ship by those on shore. Make the hawser fast about 2 feet above the tail-block. *Great care must be taken to see that there are no turns of the whip line round the hawser* and that the tally board is close to the mast (or wherever the hawser is made fast) to allow the breeches buoy to come close to it. When the hawser has been made fast on board, unbend the whip from the hawser and see that the bight of the whip has not been hitched to any part of the vessel and that it runs free in the block. Then signal to the shore as before.

The breeches buoy (a lifebuoy fitted with canvas breeches) will then be hauled off to the ship by the endless whip, travelling on the hawser. When the first person has got into the buoy, signal as before and they will be hauled ashore. As soon as they are ashore the breeches buoy will be hauled out again.

Signals	*Signification*
1. *By Day* — Vertical motion of a white flag or arms, or green star signal.	In general — "Affirmative".
By Night — Vertical motion of a white flare or light, or green star signal.	Specifically — "Rocket line is held". "Tail block is made fast". "Hawser is made fast". "Man is in the breeches buoy". "Haul away".
(Aid to memory: nodding head).	In general — "Negative".
2. *By Day* — Horizontal motion of a white flag or arms extended horizontally, or red star signal.	Specifically — "Slack away". "Avast hauling".
· *By Night* — Horizontal motion of a white light or flare, or red star signal.	
(Aid to memory: shaking head).	

Rope and Rigging — Rope is made of (1) hemp, either manila or sisal; (2) nylon terylene and other synthetic fibres; (3) steel wire.

The size of rope is determined by its diameter. A rope gauge is used to measure it.

Rope is supplied in coils of about 112 fathoms; hawsers (heavy ropes used for mooring ship) about 90 fathoms.

Cordage of less than 1 inch in circumference is known as line, ratline, pointline, etc.

Hemp Rope — Is used for general purposes, running rigging, tackles, hawsers, etc. It is usually three-stranded rope, laid up right-handed; that is to say, the strands run upwards from left to right (*see* Fig. 1). Each strand is made up of a certain number of yarns.

Right-handed rope is always coiled down right-handed; that is, in the same direction as the hands of a clock go round. Left-handed rope and log lines are coiled left-handed.

To find the breaking strain of hemp rope in tonnes, square the diameter in millimetres and divide by 150. (*e.g.* What is the breaking strain of a 20mm rope? $20 \times 20 \div 150 = 2 \cdot 66$ tonnes).

The safe working load is considered to be one-sixth of the ultimate strength, or breaking strain, of the rope for normal work.

Nylon Rope — Is almost twice as strong as hemp rope of the same size. It is much more elastic and because it stretches so, knots must be very carefully tied and more tucks are needed when splicing it.

Polythene — About as strong as hemp, is lighter than nylon and floats. It is often highly coloured.

Steel Wire Rope — Is used in rigging, mooring, etc. It is right-handed rope made up of six strands twisted round a hemp heart, each strand consisting of wires also twisted round a hemp heart. This makes it flexible. Standing rigging (stays, shrouds, etc.) need not be flexible and the strands of such wire are made entirely of wire. To find the breaking strain of flexible wire rope in tonnes, square the diameter in millimetres and divide by 25. (*e.g.* What is the breaking strain of a 20mm wire? $20 \times 20 \div 25 = 16$ tonnes). The SWL is one-sixth of the ultimate strength.

Cable-laid Rope — Is made from three ropes laid up together left-handed to form one rope.

Bolt Rope — Is tarred hemp rope with long lays, for roping sails and awnings. Sails are strengthened along their edge by bolt rope. This is always sewn on the port side of a fore-and-aft sail and the after side of a square sail which helps to prevent a sail being placed the wrong way round when being prepared for hoisting. Awnings have the bolt rope underneath.

COMPARATIVE TABLE OF APPROXIMATE
BREAKING LOADS IN TONS

Diameter mm	Steel Wire (6 strands)	Nylon (3 strands)	Manilla	Hemp
24	18·6	9	3·4	2·6
32	33·2	15	6	4·6
40	71	23	9·3	7·2
48	104·3	33	13·5	10·3
56	136·2	44	18·4	14·1
64	168·3	57	24	18·4
72		72	30·3	23·3
80		90	37·5	28·7

KNOTS

Wall Knot (Fig. 1) — Unlay the end of the rope and (1) take the left-hand strand *A* and loop it back across the front of the rope. (2) Take strand *B* and loop it over the end of strand *A*. (3) Loop strand *C* over the end of strand *B* and tuck its end into the bight formed by strand *A*. Work the knot taut. When the ends are followed round again the knot becomes a "Stopper Knot".

Crown Knot (Fig. 1) — Made in the same way as the "Wall Knot", but with the strands placed over instead of round the rope. (1) Place strand *A* over the top with the ends between the other two strands. (2) Put *B* over the end of *A*. (3) Place *C* over the end of *B* and push its end through loop formed by *A*. Work the knot taut.

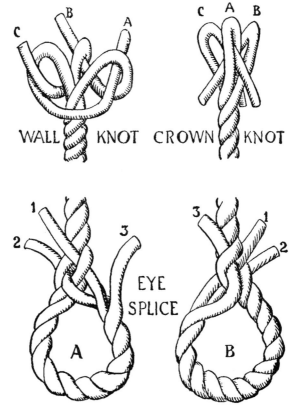

Fig. 1 — Knots and Eye-splice.

Manrope Knot — Used at the end of the manropes. (1) Make a wall. (2) Make a crown on top of it. (3) Let the strands follow the wall round and tuck them. (4) Follow the crown round. If the wall and crown are followed round once more (making three parts altogether) the knot becomes a "Turk's Head".

To make a *Turk's Head* with a single line. (1) Make a clove hitch with one very long end. (2) Holding the hitch with the long end away from you, cross the two parts beyond the long end right over left. (3) Tuck the long end under the right one and up between them. (4) Carry the long end over the left one, under the part ahead of it and up the centre. The short end will be found to come out here also. (5) Pull the short end out so that it hangs on the right, and place the long end over the right-hand part and push it through where the short end now comes out. The knot is now formed and the long end must be passed twice round again following the part it lies alongside all the time.

SPLICES

(Eye-splice (Hemp) Fig. 1 — Unlay the end of the rope and bend the rope to the size required. Spread out the three strands and (1) tuck the middle one under a strand in the rope, (2) tuck the left hand one under the strand to the left of the first strand. (3) Turn the splice round (Fig. 1B) and tuck the third strand from right to left under the strand not already used. Tuck all the strands twice more. To finish the splice, if the rope is a hawser, divide the strands and seize the halves of one strand to the halves of the next. If the splice is to be served, taper it off by dividing the strands and tucking half, then cut off the ends. Remember when splicing hemp rope that all strands are tucked from right to left, over one and under one, and no two strands should come out at the same place.

(Eye-splice (Wire) — Put a good stout whipping some distance from the end of the rope, whip each strand and unlay them up to the whipping. Cut out the hemp hearts from strands and rope. Bend the wire round to form the eye and seize it in position so that the strands lie on top of the rope in order. It is a good plan now to hang the rope up by its eye. There are several methods of tucking the strands. Thrust the marline-spike through the centre of the rope, *i.e.,* beneath three strands (avoiding the heart) and tuck the top strand. Withdraw the spike and thrust it again beneath two strands, tucking the second strand so that it goes in at the same place as the first, but passes under only two strands. Enter the third strand in the same place as the other two, but tuck it under one strand only. Tuck each of the remaining three strands under one. The eye is now formed and all that remains to be done is to tuck each strand four times round the strand under which it has already been tucked, working the marline-spike down the rope and taking long lays so that the strands lie snugly. The strands are now divided and one half of each is tucked twice more to taper the splice. The strands hammered gently down, the ends cut off and the splice parcelled and served.

MAKING AN EYE-SPLICE IN 'SQUARELINE' ROPE

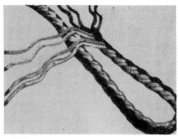

1 Rope in position for splicing. *In these photographs, left hand strands have been dyed black and right hand strands white — for demonstration purposes only.*

2 Separate strands in pairs so that white strands are ready for first tuck.

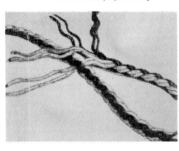

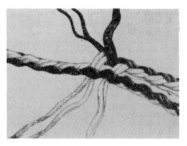

3 First pair of white strands tucked under black strands in the same direction as white strands in whole portion of rope.

4 Second pair of white strands tucked in same manner.

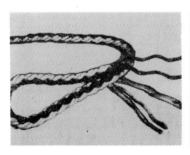

5 Turn rope over in preparation for tucking of black strands.

6 First pair of black strands each tucked separately under white strands . . .

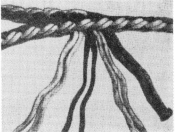

7 . . . to follow black strands in whole rope.

8 Draw the black strands as tightly as possible.

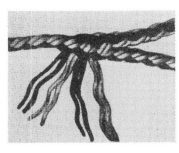

9 Turn rope over again and repeat stages 2 to 8.

10 From now on the same procedure is followed, but one strand only from each pair is tucked.
Above one of black pair is tucked under pair of white strands.

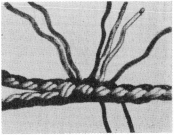

11 Second black strand is tucked under next pair of white strands.

12 Draw each black strand as tightly as possible.

Rope & Rigging 108

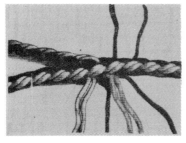

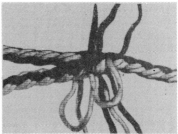

13 Turn rope over ready for tucking white strands singly.

14 One white strand from pair on right in stage 13 is tucked under pair of black strands. The other white strands from pair on left in stage 13 is taken under the first pair of black strands which were tucked in 6 and 7.

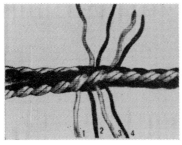

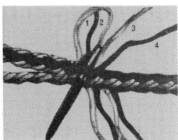

15 Draw both white strands as tightly as possible. From now on the strands numbered 1, 2, 3, 4 which remain untucked in stages 10 to 14 are used in final stages.

16 Turn rope over and tuck white strand No. 1 back under black strands which were tucked in stage 11.

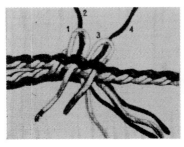

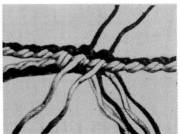

17 Tuck next white strand No. 3 under next set of tucked black strands.

18 Draw white strands tight. Repeat stages 16 and 17 with black strands and finish off.

Short Splice — Used to join two ropes together which have not to go through a block. Unlay the ends of the rope and "marry" them, putting the strands of one rope between the strands of the other. Then, working one end at a time, tuck all the strands over one and under one, from right to left as in the eye-splice. Tuck the other end in the same way. Then tuck each end again and finish off in the same way as the eye-splice.

Long Splice — For joining two ropes together which have to pass through a block. Unlay the ends of the rope to the length of $5\frac{1}{2}$ times the size of the rope, taking care to keep the turns in the strands. Marry the ends as in a short splice. Now unlay one strand from one rope, as you unlay it lay up a strand from the other rope in its place. Turn round and unlay a strand from the other rope filling it up with a strand in its place. Take a third out of all the strands. Knot the opposite strands together, heaving them well into place. Tuck all the strands once, divide them and tuck one half again.

Back Splice — Sometimes used to prevent the end of a small rope from unlaying. Unlay the end, crown it and tuck the strands twice back along the standing part.

Grummet — A rope ring made with one strand. Unlay and cut off a strand, $3\frac{1}{2}$ times the circumference of the grummet, taking care to keep the turns in. Take the strand by the middle and form a ring the size required. Lay up the first one end and then the other right round. When they meet the grummet is complete and should be finished like a long splice.

WHIPPINGS

A whipping is put on the end of a rope to prevent it unravelling.

Common Whipping — Hold the rope in the left hand with the end to the right. Lay the end of a length of twine on the rope with its end towards the rope's end and then bind the twine round both the rope and the end of the twine, working against the lay of the rope and towards the rope's end. When nearly enough turns have been put on, lay the end of the twine back along the rope and bind the last two or three turns over it. Pull the end taut and the whipping is secured.

American Whipping — Made like Common Whipping but the first end of twine is left out between the first and second half of the turns. The two ends are then knotted together.

West Country Whipping — Middle the twine and make an overhand knot round the rope. Pass the ends of the twine round the back of the rope and make another knot. Repeat this on each side of the rope alternately, and finish the whipping with a reef knot.

Palm and Needle or Sailmaker's Whipping — A more permanent whipping than those just described. Thread the twine through a needle and thrust the needle through a strand and bind the rope's end as before. When enough turns have been taken finish the whipping off by passing the needle between two strands and hauling taut. Then pass the needle between two strands at the beginning of the whipping and haul taut. Take the twine back to the end of the whipping, again pass it between two strands and finally back once more to the beginning of the whipping, so that the twine now lies across the whipping along each lay of the rope, and secure the end.

Worming, Parcelling and Serving — Rigging is often protected from wet or chafe by being wormed, parcelled and served.

To worm a rope, twist spunyard round in the lays, making the surface of the rope as smooth and round as possible.

Parcel the rope by wrapping strips of old tarred or greased canvas round the rope like a bandage, each turn just overlapping the last.

Finally serve the rope with spunyarn, binding it on tightly with a serving mallet, against the lay of the rope.

> "Worm and parcel with the lay
> Turn and serve the other way".

Eye-splices in small wire rope can be protected by parcelling them with adhesive tape.

Seizings are used to binds two parts of a rope together.

A Round Seizing (Fig. 2) — Is used when the strain on the two ropes is the same (as in the strop of a block). Put an eye in one end of the seizing stuff. Place the seizing round both parts of the rope and heave it taut. Pass the seizing round and round the two ropes, heaving each turn well taut as you go. When enough turns have been taken, reeve the end back through the turns and up through the eye of the seizing. Now pass the seizing round again over the turns already taken, each upper riding in the lay of the lower ones. When the last turn has been taken, pass the end between the last two lower turns and heave taut. Then take a round turn across the whole seizing between the two ropes, then form a clove hitch, one part lying on each side of the round turn, and secure the end by crowning it and making a wall underneath the crown.

A Flat Seizing — Is a light seizing made in the same way as a round seizing but without the upper turns.

A Racking Seizing — Is used when the strain is on one part of the rope only (as in the turnback of a shroud). Begin as in a round seizing but, instead of passing the seizing right round the ropes, pass it over and under in a figure-of-eight. When sufficient racking turns have been taken in this way, pass the end up between the ropes and then work back over the top of the racking turns, taking the seizing right round the rope as in a round seizing. Each roundabout turn should lie in the lays of the racking turns. Finish the seizing off with a clove hitch as in the round seizing.

Blocks and Tackles (pronounce "Taykuls").

The parts of a block are shown in Fig. 2. Blocks are either stropped with a rope strop (as in the hook block) or internally bound with an iron strop (as the blocks of the luff tackle) (Fig. 2). Blocks are measured by the length of their shell. A hemp rope block should be three times as big as the rope it takes. Wire rope blocks should be six times the size of the rope because wire is not so flexible as hemp. Blocks used in boats and yachts are usually made of light, laminated plastic, called "tufnol".

A Gin Block — Is an iron block with a metal sheave used for cargo whips.

A Snatch Block — Is an iron-bound block with a hinge at its swallow to allow any part of the rope to be put in the block without reeving its end through first.

A Tail Block — Is a block with a rope tail spliced to its strop instead of a hook or shackle.

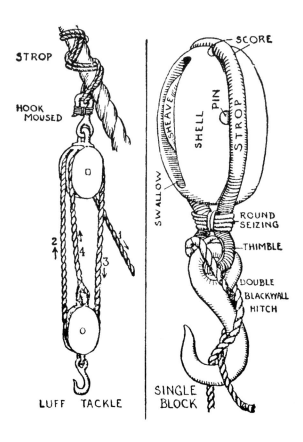

Fig. 2 — Luff Tackle and Hook Block.

Tackles or Purchases are used to obtain mechanical advantage.

Gun Tackle — A purchase with two single blocks. Mechanical advantage gained twice.

Luff Tackle (Fig. 2) — A purchase with one single and one double block. Advantage gained three or four times according to which is the movable block.

Double Luff or Twofold — A purchase with two double blocks. Mechanical advantage gained, four times.

Threefold — A purchase with two threefold blocks. Advantage gained, six times.

Fourfold — A purchase with two fourfold blocks. Advantage gained, eight times.

Handy Billy — A small tackle with a rope tail spliced to its upper block. It is sometimes known as a jigger.

Watch Tackle — A handy luff tackle used for general purposes.

Reeving a Threefold Purchase — In order to bring the greatest stress on the central part of the blocks it is usual to reeve a threefold purchase with the hauling and standing parts rove through the middle sheaves. To do so, place the blocks a convenient distance apart with the block (*A*), which is to have the hauling part, with its sheaves vertical and the other block (*B*) on its side with its sheaves horizontal. (1) Reeve the end downwards through the middle sheave of block *A*; (2) Next from right to left through the lowest sheave in block *B*; (3) Next up through the left-hand sheave in *A*; (4) Next left to right through the top sheave of *B*; (5) Next down the right-hand sheave of *A*; (6) Next right to left through the middle sheave of *B*; (7) Finally secure to block *A*.

Masts — Are nowadays usually built of steel from truck to heel, lowermast and topmast in one, but sometimes (to enable ships to pass under bridges, etc.) the topmast is fitted so that it can be "struck" or lowered. In this case, the topmast is made of wood and either "steps" or fits into the head of the lowermast (being lowered inside it) when it is called a telescopic topmast, or else it is fitted on the forepart of the mast, passing through a cap fitted to the lowermast head with its heel resting between two brackets rivetted a few feet below the lowermast cap called the trestle trees.

The weight of the topmast is borne by an iron bar called a fid, which passes through the heel of the topmast and rests upon the trestle trees.

Lowermasts are often fitted with a small platform known as the top, at the trestle trees. In merchant ships the top may carry a crow's nest and the derrick topping-lift blocks, in warships' radar aerials.

Masts of yachts are, like steel masts, hollow, but constructed of aluminium alloy or wood. Built-up wooden masts are stronger, as well as lighter, than solid spars.

Standing Rigging — Masts are supported by wire standing rigging. The rigging which supports the lowermast in an athwartship direction is known as **shrouds**; that which supports it in a fore-and-aft direction as **stays**. The lower ends of the rigging are fitted to **rigging screws**, or **turn buckles**, to enable the rigging to be set up taut. The shrouds are sometimes rattled down with **ratline** or **battens** so that men may go aloft. Seized to the bottom of the shrouds is the **sheerpole**, which keeps them from twisting.

The topmast rigging consists of topmast stays and **backstays**, which lead down on deck and set up abaft the shrouds. When backstays are rigged so that the mainsail would chafe against them when sailing free, the lee one has to be slackened and the weather one kept taut. They are then known as **running backstays**, or **runners**. Occasionally topmast shrouds are also fitted which set up in the top and carry ratlines.

Yachts with lofty masts need **spreaders** to extend the backstays outwards and widen the angle the stays make with the mast.

Funnels are supported by funnel-guys set up at equal distances round them.

A stay known as the **jumper** or **triatic stay** is fitted between foremast and funnel or fore- and mainmast to carry signal halyards. A small signal yard is sometimes crossed on the foremast for a similar purpose.

Round up — To haul through the parts of a purchase so as to bring the two blocks together.

Royal Marines — The corps of soldiers who chiefly serve aboard Her Majesty's ships.

Royal Naval Reserve and **Women's Royal Naval Reserve** — Volunteers engaged for spare-time training and receiving pay. Training is carried out afloat and at Training Centres at Belfast, Bristol, Cardiff, Dundee, Edinburgh, Glasgow, Hove, Liverpool, London and Southampton.

Run — The curve of the after part of a vessel's hull.

RULE OF THE ROAD

In order to prevent collisions, all vessels follow certain rules, make certain signals and carry certain lights. These rules are laid down in the International Regulations for Preventing Collisions at Sea. They are printed in full at the end of the chapter.

LIGHTS

A power-driven vessel when under way shall carry:

(1) *Masthead Lights* — A white light on the foremast, visible from right ahead to 2 points abaft the beam on either side for not less than 6 miles, and a similar light on the mainmast, at least 4·5 metres higher than the foremast light (vessels less than 50 metres long need not carry the mainmast light).

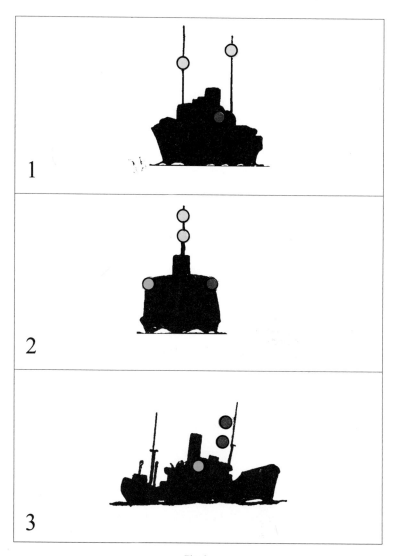

Fig. 3.

1. A power-driven vessel under way.
2. A power-driven vessel under way, end on, or, a power-driven vessel towing another vessel.
3. A vessel not under command, but making way.

Fig. 4.

4. A vessel engaged in fishing with nets or lines extending less than 150 metres, not making way through the water.
5. A pilot vessel on her station and not at anchor.
6. A vessel over 50 metres long at anchor.
   If the reader was in a steam vessel under way, he would be obliged to keep out of the way of all vessels shown in situations 1 to 6 if risk of collision existed.

(2) *Sidelights* — On the starboard side a green light and on the port side a red light visible from right ahead to 2 points abaft the beam on the starboard and port side respectively for not less than 3 miles.

(3) *Stern Light* — A white light at the stern, visible from right aft to 2 points abaft the beam on each side for not less than 3 miles. Vessels less than 7 metres long and 7 knots speed may carry a single all-round white light.

A power-driven vessel when towing another vessel shall carry the usual steaming lights, and in addition, a second white light 2 metres above or below the foremast light and shall carry a third white light 2 metres above or below the others, if the length of the tow exceeds 200 metres and a yellow towing light above the stern light.

A vessel which is not under command shall carry 2 red lights, one over the other not less than 2 metres apart and visible all round the horizon for at least 3 miles. If she is making way through the water she shall carry her sidelights and stern light, but if not she shall not carry them.

By day, she shall carry in place of the red lights 2 black balls or shapes.

A vessel restricted in her ability to manoeuvre shall carry 3 all-round lights, the highest and lowest red and the middle light white. By day, in place of the lights, 3 shapes; a diamond between two balls.

"Restricted vessels" include: Vessels working on navigational marks, pipelines and cables; dredging; surveying; transferring persons or material while under way; launching or recovering aircraft; towing unmanoeuvrable vessels.

Dredgers exhibit, in addition, 2 all-round red lights or 2 balls, vertically on the side on which an obstruction exists, 2 all-round green lights or 2 diamonds, on the side on which vessels can pass.

Deep draughted vessels, confined to a channel, exhibit 3 all-round red lights in addition to their ordinary lights, or a cylinder.

These lights and shapes are to be taken by other vessels as signals that the vessel showing them is not under command and cannot, therefore, get out of the way. They are *not* signals of distress or requiring assistance.

A sailing vessel under way and any **vessel being towed** shall carry the same lights as a steam vessel under way, with the exception of the white masthead lights which they shall never carry. Any sailing vessel may exhibit near the masthead two extra lights, the upper red and the lower green, both all-round.

Rowing boats whether under oars or sail shall have ready a lighted lantern or an electric torch which shall be shown in sufficient time to prevent collision.

Power-driven pilot vessels when on their station on pilotage duty shall carry a white light 2 metres above a red light, visible all-round the horizon for at least 3 miles, and shall also show a bright intermittent light at short intervals. When not at anchor she shall carry her sidelights and stern light.

Vessels and boats engaged in fishing. (All lights mentioned shall be visible for at least 3 miles.

With trolling (towing) lines — Lights for a power or sailing vessel under way.

Trawling — An all-round green light over a white light and, if moving through the water sidelights and stern light. A masthead light may also be carried.

Other Methods — An all-round red light over a white light and, if moving through the water, sidelights and sternlight. If the outlying gear extends more than 150 metres an additional white light, lower than the first white light and in the direction of the gear.

All vessels fishing may use a flare-up light or shine a light in the direction of their gear to attract the attention of an approaching vessel.

By Day — Vessels fishing shall display two black cones, points together. If the gear extends more than 150 metres, an additional black cone, point up, shall be displayed in the direction of the gear. *(Some additional signals are permitted).*

A vessel at anchor shall carry forward an all-round white light visible at least 3 miles, and near the stern another such light at least 4·5 metres lower than the forward one. A vessel under 50 metres in length need only show the forward light. By day a vessel at anchor shall carry a black ball or shape in place of the forward anchor light.

A vessel (except small vessels) **aground** shall carry the above lights and the two red lights of a vessel not under command, and by day, 3 black balls one over the other.

A vessel **under sail, and also propelled by machinery,** shall carry in daytime a black conical shape point down.

Hovercraft off the British coast show an amber light flashing 60 times per minute, all round the horizon for 5 miles, in addition to ordinary steaming lights.

SOUND SIGNALS

All signals prescribed for vessels under way shall be given:

By power-driven vessels upon the whistle or siren.

By sailing vessels and vessels being towed on the whistle or some other means of making the signal. There is no longer a requirement to carry a foghorn. A "prolonged blast" is from 4 to 6 seconds duration; a "short blast" about 1 second's duration.

In thick weather:

A power-driven vessel moving through the water shall sound at intervals of not more than 2 minutes 1 prolonged blast. If she is under way but stopped and not moving through the water she shall sound at intervals of not more than two minutes 2 prolonged blasts with an interval of about 1 second between them.

A sailing vessel under way shall sound at intervals of not more than 2 minutes "D" in all cases.

A vessel at anchor shall at intervals of not more than 1 minute ring the bell rapidly for about 5 seconds. A vessel over 350 feet long in addition beats a gong in the after part of the vessel. She may also sound "R".

A vessel fishing; both at anchor or underway, a vessel when **towing**, or a vessel under way which cannot get out of the way of an approaching vessel through being not under command or is **unable to manoeuvre**, shall at intervals of not more than 2 minutes sound "D". A vessel towed shall sound "B". Every 2 minutes.

A vessel aground shall make the signals for a vessel at anchor and in addition give 3 separate and distinct strokes on the bell before and after each signal on the bell.

Vessels and boats of less than 12 metres are not obliged to make the above-mentioned signals, but if they do not they shall make some other efficient sound signal at intervals of not more than 2 minutes.

Pilot vessels may sound "H" as an identity signal.

Manoeuvring and warning signals:

When vessels are in sight of one another, a power driven vessel under way in taking any course authorised or required by the Rules, shall indicate that course by the following signals, *viz:*

1 short blast to mean, "I am directing my course to starboard".

2 short blasts to mean, "I am directing my course to port".

3 short blasts to mean, "My engines are going astern".

(*see* International Code Letter Signals D, R, E, I and S, pages 175 and 176).

If a power-driven vessel, which by these Rules is to keep her course and speed, is in doubt whether sufficient action is being taken by the giving way vessel, she may make at least 5 short and rapid blasts on her whistle. These sound signals may be supplemented by similar light signals. Overtaking signals in narrow channels.

G: "I intend to overtake you and pass you on your starboard side".

Z: "I intend to overtake you and pass you on your port side".

C: "I agree to be overtaken".

STEERING AND SAILING RULES

These rules are contained in Rules 4-19 and should be very carefully read.

Risks of collision can, when circumstances permit, be ascertained by watching the bearing of the approaching vessel. If the bearing does not appreciably change, such risk should be deemed to exist. (*See* Fig. 5).

When two vessels of different types are approaching the vessel that is more easily manoeuvred usually keeps out of the way before a close quarters situation develops. When a very fast vessel is approaching a

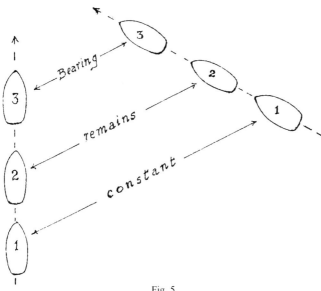

Fig. 5.

much slower vessel, the high-speed vessel will probably keep out of the way because whatever action the slow one may take will have little effect. The vessel which has to keep out of the way should not cross ahead of the other, and if she is a power-driven vessel she should, if necessary, slacken her speed, stop or reverse. She should also alter course in good time and to such an extent that the other vessel can have no doubt as to what she intends to do. A vessel which has not to give way should keep her course and speed.

Vessels and boats engaged in fishing and small craft generally should, in narrow channels, keep out of the way of large ships who may not have sufficient room to manoeuvre.

INTERNATIONAL REGULATIONS

FOR

PREVENTING COLLISIONS AT SEA

(as amended by Resolutions A464(XII), A626(15), A678(16) and A736(18))
Amended version formally adopted by the United Kingdom from 1st May 1996,
applicable internationally from 4th November, 1995.

―――

PART A. GENERAL

RULE 1

Application

(*a*) These Rules shall apply to all vessels upon the high seas and in all waters connected therewith navigable by seagoing vessels.

(*b*) Nothing in these Rules shall interfere with the operation of special rules made by an appropriate authority for roadsteads, harbours, rivers, lakes or inland waterways connected with the high seas and navigable by seagong vessels. Such special rules shall conform as closely as possible to these Rules.

(*c*) Nothing in these Rules shall interfere with the operation of any special rules made by the Government of any State with respect to additional station or signal lights, shapes or whistle signals for ships of war and vessels proceeding under convoy, or with respect to additional station or signal lights or shapes for fishing vessels engaged in fishing as a fleet. These additional station or signal lights, shapes or whistle signals shall, so far as possible, be such that they cannot be mistaken for any light, shape or signal authorized elsewhere under these Rules.

(*d*) Traffic separation schemes may be adopted by the Organization for the purpose of these Rules.

(*e*) Whenever the Government concerned shall have determined that a vessel of special construction or purpose cannot comply fully with the provisions of any of these Rules with respect to the number, position, range or arc of visibility of lights or shapes, as well as to the disposition and characteristics of sound-signalling appliances, such vessel shall comply with such other provisions in regard to the number, position, range or arc of visibility of lights or shapes, as well as to the disposition and characteristics of sound-signalling appliances, as her Government shall have determined to be the closest possible compliance with these Rules in respect of that vessel.

RULE 2

Responsibility

(*a*) Nothing in these rules shall exonerate any vessel, or the owner, master or crew thereof, from the consequences of any neglect to comply with these Rules or of the neglect of any precaution which may be required by the ordinary practice of seamen, or by the special circumstances of the case.

(*b*) In construing and complying with these Rules due regard shall be had to all dangers of navigation and collision and to any special circumstances, including the limitations of the vessels involved, which may make a departure from these rules necessary to avoid immediate danger.

RULE 3

General definitions

For the purpose of these Rules, except where the context otherwise requires.

(*a*) The word "vessel" includes every description of water craft, including non-displacement craft and seaplanes, used or capable of being used as a means of transportation on water.

(*b*) The term "power-driven vessel" means any vessel propelled by machinery.

(*c*) The term "sailing vessel" means any vessel under sail provided that propelling machinery, if fitted, is not being used.

(*d*) The term "vessel engaged in fishing" means any vessel fishing with nets, lines, trawls or other fishing apparatus which restrict manoeuvrability, but does not include a vessel fishing with trolling lines or other fishing apparatus which do not restrict manoeuvrability.

(*e*) The word "seaplane" includes any aircraft designed to manoeuvre on the water.

(*f*) The term "vessel not under command" means a vessel which through some exceptional circumstance is unable to manoeuvre as required by these Rules and is therefore unable to keep out of the way of another vessel.

(*g*) The term "vessel restricted in her ability to manoeuvre" means a vessel which from the nature of her work is restricted in her ability to manoeuvre as required by these Rules and is therefore unable to keep out of the way of another vessel.

The term "vessels restricted in their ability to manoeuvre" shall include but not be limited to:

  (i) a vessel engaged in laying, servicing or picking up a navigation mark, submarine cable or pipeline;

  (ii) a vessel engaged in dredging, surveying or underwater operations;

  (iii) a vessel engaged in replenishment or transferring persons, provisions or cargo while underway;

  (iv) a vessel engaged in the launching or recovery of aircraft;

  (v) a vessel engaged in mineclearance operations;

  (vi) a vessel engaged in a towing operation such as severely restricts the towing vessel and her tow in their ability to deviate from their course.

(*h*) The term "vessel constrained by her draught" means a power-driven vessel which because of her draught in relation to the available depth and width of navigable water is severely restricted in her ability to deviate from the course she is following.

(*i*) The word "underway" means that a vessel is not at anchor, or made fast to the shore, or aground.

(*j*) The words "length" and "breadth" of a vessel mean her length overall and greatest breadth.

(*k*) Vessels shall be deemed to be in sight of one another only when one can be observed visually from the other.

(*l*) The term "restricted visibility" means any condition in which visibility is restricted by fog, mist, falling snow, heavy rainstorms, sandstorms or any other similar causes.

PART B—STEERING AND SAILING RULES

Section I. Conduct of vessels in any condition of visibility

RULE 4

Application

Rules in this Section apply in any condition of visibility.

RULE 5

Look-out

Every vessel shall at all times maintain a proper look-out by sight and hearing as well as by all available means appropriate in the prevailing circumstances and conditions so as to make a full appraisal of the situation and the risk of collision.

RULE 6

Safe speed

Every vessel shall at all times proceed at a safe speed so that she can take proper and effective action to avoid collision and be stopped within a distance appropriate to the prevailing circumstances and conditions.

In determining a safe speed the following factors shall be among those taken into account.

(*a*) By all vessels:

    (i) the state of visibility;

    (ii) the traffic density including concentrations of fishing vessels or any other vessels;

    (iii) the manoeuvrability of the vessel with special reference to stopping distance and turning ability in the prevailing conditions;

    (iv) at night the presence of background light such as from shore lights or from back scatter of her own lights;

    (v) the state of wind, sea and current, and the proximity of navigational hazards;

    (vi) the draught in relation to the availabe depth of water.

(*b*) Additionally, by vessels with operational radar:

    (i) the characteristics, efficiency and limitations of the radar equipment;

(ii) any constraints imposed by the radar range scale in use;

(iii) the effect on radar detection of the sea state, weather and other sources of interference;

(iv) the possibility that small vessels, ice and other floating objects may not be detected by radar at an adequate range;

(v) the number, location and movement of vessels detected by radar;

(vi) the more exact assessment of the visibility that may be possible when radar is used to determine the range of vessels or other objects in the vicinity.

RULE 7

Risk of collision

(*a*) Every vessel shall use all available means appropriate to the prevailing circumstances and conditions to determine if risk of collision exists. If there is any doubt such risk shall be deemed to exist.

(*b*) Proper use shall be made of radar equipment if fitted and operational, including long-range scanning to obtain early warning of risk of collision and radar plotting or equivalent systematic observation of detected objects.

(*c*) Assumptions shall not be made on the basis of scanty information, especially scanty radar information.

(*d*) In determining if risk of collision exists the following considerations shall be among those taken into account:

(i) such risk shall be deemed to exist if the compass bearing of an approaching vessel does not appreciably change;

(ii) such risk may sometimes exist even when an appreciable bearing change is evident, particularly when approaching a very large vessel or a tow or when approaching a vessel at close range.

RULE 8

Action to avoid collision

(*a*) Any action taken to avoid collision shall, if the circumstances of the case admit, be positive, made in ample time and with due regard to the observance of good seamanship.

(*b*) Any alteration of course and/or speed to avoid collision shall, if the circumstances of the case admit, be large enough to be readily apparent to another vessel observing visually or by radar; a succession of small alterations of course and/or speed should be avoided.

(*c*) If there is sufficient sea room, alteration of course alone may be the most effective action to avoid a close-quarters situation provided that it is made in good time, is substantial and does not result in another close-quarters situation.

(*d*) Action taken to avoid collision with another vessel shall be such as to result in passing at a safe distance. The effectiveness of the action shall be carefully checked until the other vessel is finally past and clear.

(*e*) If necessary to avoid collision or allow more time to assess the situation, a vessel shall slacken her speed or take all way off by stopping or reversing her means of propulsion.

(*f*) (i) A vessel which by any of these Rules is required not to impede the passage or safe passage of another vessel shall, when required by the circumstances of the case, take early action to allow sufficient sea room for the safe passage of the other vessel.

(ii) A vessel required not to impede the passage or safe passage of another vessel is not relieved of this obligation if approaching the other vessel so as to involve risk of collision and shall, when taking action, have full regard to the action which may be required by the Rules of this Part.

(iii) A vessel whose passage is not to be impeded remains fully obliged to comply with the Rules of this Part when the two vessels are approaching one another so as to involve risk of collision.

RULE 9

Narrow channels

(*a*) A vessel proceeding along the course of a narrow channel or fairway shall keep as near to the outer limit of the channel or fairway which lies on her starboard side as is safe and practicable.

(*b*) A vessel of less than 20 metres in length or a sailing vessel shall not impede the passage of a vessel which can safely navigate only within a narrow channel or fairway.

(*c*) A vessel engaged in fishing shall not impede the passage of any other vessel navigating within a narrow channel or fairway.

(*d*) A vessel shall not cross a narrow channel or fairway if such crossing impedes the passage of a vessel which can safely navigate only within such channel or fairway. The latter vessel may use the sound signal prescribed in Rule 34 (*d*) if in doubt as to the intention of the crossing vessel.

(*e*) (i) In a narrow channel or fairway when overtaking can take place only if the vessel to be overtaken has to take action to permit safe passing, the vessel intending to overtake shall indicate her intention by sounding the appropriate signal prescribed in Rule 34 (*c*) (i). The vessel to be overtaken shall, if in agreement, sound the appropriate signal prescribed in Rule 34 (*c*) (ii) and take steps to permit safe passing. If in doubt she may sound the signals prescribed in Rule 34 (*d*).

(ii) This rule does not relieve the overtaking vessel of her obligation under Rule 13.

(*f*) A vessel nearing a bend or an area of a narrow channel or fairway where other vessels may be obscured by an intervening obstruction shall navigate with particular alertness and caution and shall sound the appropriate signal prescribed in Rule 34 (*e*).

(*g*) Any vessel shall, if the circumstances of the case admit, avoid anchoring in a narrow channel.

RULE 10

Traffic separation schemes

(*a*) This Rule applies to traffic separation schemes adopted by the Organization and does not relieve any vessel of her obligation under any other Rule.

(*b*) A vessel using a traffic separation scheme shall:

  (i) proceed in the appropriate traffic lane in the general direction of traffic flow for that lane;

  (ii) so far as practicable keep clear of a traffic separation line or separation zone;

  (iii) normally join or leave a traffic lane at the termination of the lane, but when joining or leaving from either side shall do so at as small an angle to the general direction of traffic flow as practicable.

(*c*) A vessel shall, so far as practicable, avoid crossing traffic lanes but if obliged to do so shall cross on a heading as nearly as practicable at right angles to the general direction of traffic flow.

(*d*) (i) A vessel shall not use an inshore traffic zone when she can safely use the appropriate traffic lane within the adjacent traffic separation scheme. However, vessels of less than 20 metres in length, sailing vessels and vessels engaged in fishing may use the inshore traffic zone.

  (ii) Not withstanding subparagraph (*d*) (i), a vessel may use an inshore traffic zone when en route to or from a port, offshore installation or structure, pilot station or any other place situated within the inshore traffic zone or to avoid immediate danger.

(*e*) A vessel other than a crossing vessel or a vessel joining or leaving a lane shall not normally enter a separation zone or cross a separation line except:

  (i) In cases of emergency to avoid immediate danger;

  (ii) to engage in fishing within a separation zone.

(*f*) A vessel navigating in areas near the terminations of traffic separation schemes shall do so with particular caution.

(*g*) A vessel shall so far as practicable avoid anchoring in a traffic separation scheme or in areas near its terminations.

(*h*) A vessel not using a traffic separation scheme shall avoid it by as wide a margin as is practicable.

(*i*) A vessel engaged in fishing shall not impede the passage of any vessel following the traffic lane.

(*j*) A vessel of less than 20 metres in length or a sailing vessel shall not impede the safe passage of a power-driven vessel following a traffic lane.

(*k*) A vessel restricted in her ability to manoeuvre when engaged in an operation for the maintenance of safety of navigation in a traffic separation scheme is exempted from complying with this Rule to the extent necessary to carry out the operation.

(*l*) A vessel restricted in her ability to manoeuvre when engaged in an operation for the laying, servicing or picking up of a submarine cable, within a traffic separation scheme, is exempted from complying with this Rule to the extent necessary to carry out the operation.

Section II. Conduct of vessels in sight of one another

RULE 11

Application

Rules in this Section apply to vessels in sight of one another.

RULE 12

Sailing vessels

(*a*) When two sailing vessels are approaching one another, so as to involve risk of collision, one of them shall keep out of the way of the other as follows:

(i) when each has the wind on a different side, the vessel which has the wind on the port side shall keep out of the way of the other;

(ii) when both have the wind on the same side, the vessel which is to windward shall keep out of the way of the vessel which is to leeward;

(iii) if a vessel with the wind on the port side sees a vessel to windward and cannot determine with certainty whether the other vessel has the wind on the port or on the starboard side, she shall keep out of the way of the other.

(*b*) For the purposes of this Rule the windward side shall be deemed to be the side opposite to that on which the mainsail is carried or, in the case of a square-rigged vessel, the side opposite to that on which the largest fore-and-aft sail is carried.

RULE 13

Overtaking

(*a*) Notwithstanding anything contained in the Rules of Part B, Sections I and II, any vessel overtaking any other shall keep out of the way of the vessel being overtaken.

(*b*) A vessel shall be deemed to be overtaking when coming up with another vessel from a direction more than 22·5 degrees abaft her beam, that is, in such a position with reference to the vessel she is overtaking, that at night she would be able to see only the sternlight of that vessel but neither of her sidelights.

(*c*) When a vessel is in doubt as to whether she is overtaking another, she shall assume that this is the case and act accordingly.

(*d*) Any subsequent alteration of the bearing between the two vessels shall not make the overtaking vessel a crossing vessel within the meaning of these Rules or relieve her of the duty of keeping clear of the overtaken vessel until she is finally past and clear.

RULE 14

Head-on situation

(*a*) When two power-driven vessels are meeting on reciprocal or nearly reciprocal courses so as to involve risk of collision each shall alter her course to starboard so that each shall pass on the port side of the other.

(*b*) Such a situation shall be deemed to exist when a vessel sees the other ahead or nearly ahead and by night she could see the masthead lights of the other in a line or nearly in a line and/or both sidelights and by day she observes the corresponding aspect of the other vessel.

(*c*) When a vessel is in any doubt as to whether such a situation exists she shall assume that it does exist and act accordingly.

RULE 15

Crossing situation

When two power-driven vessels are crossing so as to involve risk of collision, the vessel which has the other on her own starboard side shall keep out of the way and shall, if the circumstances of the case admit, avoid crossing ahead of the other vessel.

RULE 16

Action by give-way vessel

Every vessel which is directed to keep out of the way of another vessel shall, so far as possible, take early and substantial action to keep well clear.

RULE 17

Action by stand-on vessel

(*a*) (i) Where one of the two vessels is to keep out of the way the other shall keep her course and speed.

(ii) The latter vessel may however take action to avoid collision by her manoeuvre alone, as soon as it becomes apparent to her that the vessel required to keep out of the way is not taking appropriate action in compliance with these Rules.

(*b*) When, from any cause, the vessel required to keep her course and speed finds herself so close that collision cannot be avoided by the action of the give-way vessel alone, she shall take such action as will best aid to avoid collision.

(*c*) A power-driven vessel which takes action in a crossing situation in accordance with sub-paragraph (*a*) (ii) of this Rule to avoid collision with another power-driven vessel shall, if the circumstances of the case admit, not alter course to port for a vessel on her own port side.

(*d*) This rule does not relieve the give-way vessel of her obligation to keep out of the way.

RULE 18

Responsibilities between vessels

Except where Rules 9, 10 and 13 otherwise require:

(*a*) A power-driven vessel underway shall keep out of the way of:
  (i) a vessel not under command;
  (ii) a vessel restricted in her ability to manoeuvre;
  (iii) a vessel engaged in fishing;
  (iv) a sailing vessel;

(*b*) A sailing vessel underway shall keep out of the way of:
  (i) a vessel not under command;
  (ii) a vessel restricted in her ability to manoeuvre;
  (iii) a vessel engaged in fishing.

(*c*) A vessel engaged in fishing when underway shall, so far as possible, keep out of the way of:
  (i) a vessel not under command;
  (ii) a vessel restricted in her ability to manoeuvre.

(*d*) (i) Any vessel other than a vessel not under command or a vessel restricted in her ability to manoeuvre shall, if the circumstances of the case admit, avoid impeding the safe passage of a vessel constrained by her draught, exhibiting the signals in Rule 28.

(ii) A vessel constrained by her draught shall navigate with particular caution having full regard to her special condition.

(*e*) A seaplane on the water shall, in general, keep well clear of all vessels and avoid impeding their navigation. In circumstances, however, where risk of collision exists, she shall comply with the Rules of this Part.

Section III. Conduct of vessels in restricted visibility

RULE 19

Conduct of vessels in restricted visibility

(*a*) This Rule applies to vessels not in sight of one another when navigating in or near an area restricted of visibility.

(*b*) Every vessel shall proceed at a safe speed adapted to the prevailing circumstances and conditions of restricted visibility. A power-driven vessel shall have her engines ready for immediate manoeuvre.

(*c*) Every vessel shall have due regard to the prevailing circumstances and conditions of restricted visibility when complying with the Rules of Section I of this Part.

(*d*) A vessel which detects by radar alone the presence of another vessel shall determine if a close-quarters situation is developing and/or risk of collision exists. If so, she shall take avoiding action in ample time, provided that when such action consists of an alteration of course, so far as possible the following shall be avoided.

(i) an alteration of course to port for a vessel forward of the beam, other than for a vessel being overtaken;

(ii) an alteration of course towards a vessel abeam or abaft the beam.

(*e*) Except where it has been determined that a risk of collision does not exist, every vessel which hears apparently forward of her beam the fog signal of another vessel, or which cannot avoid a close-quarters situation with another vessel forward of her beam, shall reduce her speed to the minimum at which she can be kept on her course. She shall if necessary take all her way off and in any event navigate with extreme caution until danger of collision is over.

PART C—LIGHTS AND SHAPES

RULE 20

Application

(*a*) Rules in this part shall be complied with in all weathers.

(*b*) The Rules concerning lights shall be complied with from sunset to sunrise, and during such times no other lights shall be exhibited, except such lights as cannot be mistaken for the lights specified in these Rules or do not impair their visibility or distinctive character, or interfere with the keeping of a proper lookout.

(*c*) The lights prescribed by these Rules shall, if carried, also be exhibited from sunrise to sunset in restricted visibility and may be exhibited in all other circumstances when it is deemed necessary.

(*d*) The Rules concerning shapes shall be complied with by day.

(*e*) The lights and shapes specified in these Rules shall comply with the provisions of Annex I to these Regulations.

RULE 21

Definitions

(*a*) "Masthead light" means a white light placed over the fore and aft centreline of the vessel showing an unbroken light over an arc of the horizon of 225 degrees and so fixed as to show the light from right ahead to 22·5 degrees abaft the beam on either side of the vessel.

(*b*) "Sidelights" means a green light on the starboard side and a red light on the port side each showing an unbroken light over an arc of the horizon of 112·5 degrees and so fixed as to show the light from right ahead to 22·5 degrees abaft the beam on its respective side. In a vessel of less than 20 metres in length the sidelights may be combined in one lantern carried on the fore and aft centreline of the vessel.

(*c*) "Sternlight" means a white light placed as nearly as practicable at the stern showing an unbroken light over an arc of the horizon of 135 degrees and so fixed as to show the light 67·5 degrees from right aft on each side of the vessel.

(*d*) "Towing light" means a yellow light having the same characteristics as the "sternlight" defined in paragraph (*c*) of this Rule.

(*e*) "All round light" means a light showing an unbroken light over an arc of the horizon of 360 degrees.

(*f*) "Flashing light" means a light flashing at regular intervals at a frequency of 120 flashes or more per minute.

RULE 22

Visibility of lights

The lights prescribed in these Rules shall have an intensity as specified in Section 8 of Annex I to these Regulations so as to be visible at the following minimum ranges:

(*a*) In vessels of 50 metres or more in length:
  —a masthead light, 6 miles;
  —a sidelight, 3 miles;
  —a sternlight, 3 miles;
  —a towing light, 3 miles;
  —a white, red, green or yellow all-round light, 3 miles.

(*b*) In vessels of 12 metres or more in length but less than 50 metres in length;
  —a masthead light, 5 miles; except that where the length of the vessel is less than 20 metres, 3 miles;
  —a sidelight, 2 miles;
  —a sternlight, 2 miles;
  —a towing light, 2 miles;
  —a white, red, green or yellow all-round light, 2 miles.

(*c*) In vessels of less than 12 metres in length:
—a masthead light, 2 miles;
—a sidelight, 1 mile;
—a sternlight, 2 miles;
—a towing light, 2 miles;
—a white, red, green or yellow all-round light, 2 miles.

(*d*) In inconspicuous,partly submerged vessels or objects being towed:
—a white all-round light, 3 miles.

RULE 23

Power-driven vessels underway

(*a*) A power-driven vessel underway shall exhibit:
(i) a masthead light forward;
(ii) a second masthead light abaft of and higher than the forward one; except that a vessel of less than 50 metres in length shall not be obliged to exhibit such light but may do so;
(iii) sidelights;
(iv) a sternlight.

(*b*) An air-cushion vessel when operating in the non-displacement mode shall, in addition to the lights prescribed in paragraph (*a*) of this Rule, exhibit an all-round flashing yellow light.

(*c*) (i) A power-driven vessel of less than 12 metres in length may in lieu of the lights prescribed in paragraph (*a*) of this Rule exhibit an all-round white light and sidelights;
(ii) A power-driven vessel of less than 7 metres in length whose maximum speed does not exceed 7 knots may in lieu of the lights prescribed in paragraph (*a*) of this Rule exhibit an all-round white light and shall, if practicable, also exhibit sidelights;
(iii) The masthead light or all-round white light on a power-driven vessel of less than 12 metres in length may be displaced from the fore and aft centreline of the vessel if centreline fitting is not practicable, provided that the sidelights are combined in one lantern which shall be carried on the fore and aft centreline of the vessel or located as nearly as practicable in the same fore and aft line as the masthead light or the all-round white light.

RULE 24

Towing and pushing

(*a*) A power-driven vessel when towing shall exhibit:
(i) instead of the lights prescribed in Rule 23 (*a*) (i) or (*a*) (ii), two masthead lights in a vertical line. When the length of the tow, measuring from the stern of the towing vessel to the after end of the tow exceeds 200 metres, three such lights in a vertical line.
(ii) sidelights;
(iii) a sternlight;
(iv) a towing light in a vertical line above the sternlight;
(v) when the length of the tow exceeds 200 metres, a diamond shape where it can best be seen.

(*b*) When pushing a vessel and a vessel being pushed ahead are rigidly connected in a composite unit they shall be regarded as a power-driven vessel and exhibit the lights prescribed in Rule 23.

(*c*) A power-driven vessel when pushing ahead or towing alongside, except in the case of a composite unit, shall exhibit:

(i) instead of the light prescribed in Rule 23 (*a*) (i) or (*a*) (ii), two masthead lights in a vertical line;

(ii) sidelights;

(iii) a sternlight.

(*d*) A power-driven vessel to which paragraph (*a*) or (*c*) of this Rule applies shall also comply with Rule 23 (*a*) (ii).

(*e*) A vessel or object being towed, other than those mentioned in paragraph (*g*) of this Rule, shall exhibit:

(i) sidelights;

(ii) a sternlight;

(iii) when the length of the tow exceeds 200 metres, a diamond shape where it can best be seen.

(*f*) Provided that any number of vessels being towed alongside or pushed in a group shall be lighted as one vessel.

(i) a vessel being pushed ahead, not being part of a composite unit, shall exhibit at the forward end, sidelights;

(ii) a vessel being towed alongside shall exhibit a sternlight and at the forward end, sidelights.

(*g*) An inconspicuous, partly submerged vessel or object, or combination of such vessels or objects being towed, shall exhibit:

(i) if it is less than 25 metres in breadth, one all-round white light at or near the forward end and one at or near the after end except that dracones need not exhibit a light at or near the forward end;

(ii) if it is 25 metres or more in breadth, two additional all-round white lights at or near the extremities of its breadth;

(iii) if it exceeds 100 metres in length, additional all-round white lights between the lights prescribed in sub-paragraphs (i) and (ii) so that the distance between the lights shall not exceed 100 metres;

(iv) a diamond shape at or near the aftermost extremity of the last vessel or object being towed and if the length of the tow exceeds 200 metres an additional diamond shape where it can best be seen and located as far forward as is practicable.

(*h*) Where from any sufficient cause it is impracticable for a vessel or object being towed to exhibit the lights or shapes prescribed in paragraph (*e*) or (*g*) of this rule, all possible measures shall be taken to light the vessel or object towed or at least to indicate the presence of such vessel or object.

(*i*) Where from any sufficient cause it is impracticable for a vessel not normally engaged in towing operations to display the lights prescribed in paragraph (*a*) or (*c*) of this Rule, such vessel shall not be required to exhibit those lights when engaged in towing another vessel in distress or otherwise in need of assistance. All possible measures shall be taken to indicate the nature of the relationship between the towing vessel and the vessel being towed as authorized by Rule 36, in particular by illuminating the towline.

RULE 25

Sailing vessels underway and vessels under oars

(*a*) A sailing vessel underway shall exhibit:
    (i) sidelights;
    (ii) a sternlight.

(*b*) In a sailing vessel of less than 20 metres in length the lights prescribed in paragraph (*a*) of this Rule may be combined in one lantern carried at or near the top of the mast where it can best be seen.

(*c*) A sailing vessel underway may, in addition to the lights prescribed in paragraph (*a*) of this Rule, exhibit at or near the top of the mast, where they can best be seen, two all-round lights in a vertical line, the upper being red and the lower green, but these lights shall not be exhibited in conjunction with the combined lantern permitted by paragraph (*b*) of this Rule.

(*d*) (i) A sailing vessel of less than 7 metres in length shall, if practicable, exhibit the lights prescribed in paragraphs (*a*) or (*b*) of this Rule, but if she does not, she shall have ready at hand an electric torch or lighted lantern showing a white light which shall be exhibited in sufficient time to prevent collision.

    (ii) A vessel under oars may exhibit the lights prescribed in this Rule for sailing vessels, but if she does not, she shall have ready at hand an electric torch or lighted lantern showing a white light which shall be exhibited in sufficient time to prevent collision.

(*e*) A vessel proceeding under sail when also being propelled by machinery shall exhibit forward where it can best be seen a conical shape, apex downwards.

RULE 26

Fishing vessels

(*a*) A vessel engaged in fishing, whether underway or at anchor, shall exhibit only the lights and shapes prescribed in this Rule.

(*b*) A vessel when engaged in trawling, by which is meant the dragging through the water of a dredge net or other apparatus used as a fishing appliance, shall exhibit:
    (i) two all-round lights in a vertical line, the upper being green and the lower white, or a shape consisting of two cones with their apexes together in a vertical line one above the other.
    (ii) a masthead light abaft and higher than the all-round green light; a vessel of less than 50 metres in length shall not be obliged to exhibit such a light but may do so;
    (iii) when making way through the water, in addition to the lights prescribed in this paragraph, sidelights and a sternlight.

(*c*) A vessel engaged in fishing, other than trawling, shall exhibit:
    (i) two all-round lights in a vertical line, the upper being red and the lower white, or a shape consisting of two cones with apexes together in a vertical line one above the other;
    (ii) when there is outlying gear extending more than 150 metres horizontally from the vessel, an all-round white light or a cone apex upwards in the direction of the gear;
    (iii) when making way through the water, in addition to the lights prescribed in this paragraph, sidelights and a sternlight.

(*d*) The additional signals described in Annex II to these Regulations apply to a vessel engaged in fishing in close proximity to other vessels engaged in fishing.

(*e*) A vessel when not engaged in fishing shall not exhibit the lights or shapes prescribed in this Rule, but only those prescribed for a vessel of her length.

RULE 27

Vessels not under command or restricted in their ability to manoeuvre

(*a*) A vessel not under command shall exhibit:

  (i) two all-round red lights in a vertical line where they can best be seen;

 (ii) two balls or similar shapes in a vertical line where they can best be seen;

(iii) when making way through the water, in addition to the lights prescribed in this paragraph, sidelights and a sternlight.

(*b*) A vessel restricted in her ability to manoeuvre, except a vessel engaged in mineclearance operations, shall exhibit:

  (i) three all-round lights in a vertical line where they can best be seen. The highest and lowest of these lights shall be red and the middle light shall be white;

 (ii) three shapes in a vertical line where they can best be seen. The highest and lowest of these shapes shall be balls and the middle one a diamond;

(iii) when making way through the water, a masthead light or lights, sidelights and a sternlight, in addition to the lights prescribed in sub-pargraph (i);

(iv) when at anchor, in addition to the lights or shapes prescribed in sub-paragraphs (i) and (ii), the light, lights or shape prescribed in Rule 30.

(*c*) A power-driven vessel engaged in a towing operation such as severely restricts the towing vessel and her tow in their ability to deviate from their course shall, in addition to the lights or shapes prescribed in Rule 24 (*a*), exhibit the lights or shapes prescribed in sub-paragraphs (*b*) (i) and (ii) of this Rule.

(*d*) A vessel engaged in dredging or undewater operations, when restricted in her ability to manoeuvre, shall exhibit the lights and shapes prescribed in sub-paragraphs (*b*) (i), (ii) and (iii) of this Rule and shall in addition, when obstruction exists, exhibit:

  (i) two all-round red lights or two balls in a vertical line to indicate the side on which the obstruction exists;

 (ii) two all-round green lights or two diamonds in a vertical line to indicate the side on which another vessel may pass;

(iii) when at anchor, the lights or shapes prescribed in this paragraph instead of the lights or shape prescribed in Rule 30.

(*e*) Whenever the size of a vessel engaged in diving operations makes it impracticable to exhibit all lights and shapes prescribed in paragraph (*d*) of this Rule, the following shall be exhibited:

  (i) three all-round lights in a vertical line where they can best be seen. The highest and lowest of these lights shall be red and the middle light shall be white;

 (ii) a rigid replica of the International Code flag "A" not less than 1 metre in height. Measures shall be taken to ensure its all round visibility.

(*f*) A vessel engaged in mineclearance operations shall in addition to the lights prescribed for a power-driven vesssel in Rule 23 or to the lights or shape prescribed for a vessel at anchor in Rule 30 as appropriate, exhibit three all-round green lights or three balls. One of these lights or shapes shall be exhibited near the foremast head and one at each end of the fore yard. These lights or shapes indicate that it is dangerous for another vessel to approach within 1000 metres of the mineclearance vessel.

(*g*) Vessels of less than 12 metres in length, except those engaged in diving operations, shall not be required to exhibit the lights and shapes prescribed in this Rule.

(*h*) The signals prescribed in this Rule are not signals of vessels in distress and requiring assistance. Such signals are contained in Annex IV to these regulations.

RULE 28

Vessels constrained by their draught

A vessel constrained by her draught may, in addition to the lights prescribed for power-driven vessels in Rule 23, exhibit where they can best be seen three all-round red lights in a vertical line, or a cylinder.

RULE 29

Pilot vessels

(*a*) A vessel engaged on pilotage duty shall exhibit:
- (i) at or near the masthead, two all-round lights in a vertical line, the upper being white and the lower red;
- (ii) when underway, in addition, sidelights and a sternlight;
- (iii) when at anchor, in addition to the lights prescribed in sub-paragraph (i), the light, lights or shape prescribed in Rule 30 for vessels at anchor.

(*b*) A pilot vessel when not engaged on pilotage duty shall exhibit the lights or shapes prescribed for a similar vessel of her length.

RULE 30

Anchored vessels and vessels aground

(*a*) A vessel at anchor shall exhibit where it can best be seen:
- (i) in the fore part, an all-round white light or one ball;
- (ii) at or near the stern and at a lower level than the light prescribed in sub-paragraph (i), an all-round white light.

(*b*) A vessel of less than 50 metres in length may exhibit an all-round white light where it can best be seen instead of the lights prescribed in paragraph (*a*) of this Rule.

(*c*) A vessel at anchor may, and a vessel of 100 metres and more in length shall, also use the available working or equivalent lights to illuminate her decks.

(*d*) A vessel aground shall exhibit the lights prescribed in paragraphs (*a*) or (*b*) of this Rule and in addition, where they can best be seen:
- (i) two all-round red lights in a vertical line;
- (ii) three balls in a vertical line.

(*e*) A vessel of less than 7 metres in length, when at anchor, not in or near a narrow channel, fairway or anchorage, or where other vessels normally navigate, shall not be required to exhibit the lights or shapes prescribed in paragraphs (*a*) and (*b*) of this Rule.

(*f*) A vessel of less than 12 metres in length, when aground, shall not be required to exhibit the lights or shapes prescribed in sub-paragraphs (*d*) (i) and (ii) of this Rule.

RULE 31

Seaplanes

Where it is impracticable for a seaplane to exhibit lights and shapes of the characteristics or in the positions prescribed in the Rules of this Part she shall exhibit lights and shapes as closely similar in characteristics and position as is possible.

PART D—SOUND AND LIGHT SIGNALS

RULE 32

Definitions

(*a*) The word "whistle" means any sound signalling appliance capable of producing the prescribed blasts and which complies with the specifications in Annex III to these regulations.

(*b*) The term "short blast" means a blast of about one second's duration.

(*c*) The term "prolonged blast" means a blast of from four to six seconds' duration.

RULE 33

Equipment for sound signals

(*a*) A vessel of 12 metres or more in length shall be provided with a whistle and a bell and a vessel of 100 metres or more in length shall, in addition, be provided with a gong, the tone and sound of which cannot be confused with that of the bell. The whistle, bell and gong shall comply with the specifications in Annex III to these regulations. The bell or gong or both may be replaced by other equipment having the same respective sound characteristics, provided that manual sounding of the prescribed signals shall always be possible.

(*b*) A vessel of less than 12 metres in length shall not be obliged to carry the sound signalling appliances prescribed in paragraph (*a*) of this Rule but if she does not, she shall be provided with some other means of making an efficient sound signal.

RULE 34

Manoeuvring and warning signals

(*a*) When vessels are in sight of one another, a power-driven vessel underway, when manoeuvring as authorized or required by these Rules, shall indicate that manoeuvre by the following signals on her whistle:

—one short blast to mean "I am altering my course to starboard";

—two short blasts to mean "I am altering my course to port";

—three short blasts to mean "I am operating astern propulsion".

(*b*) Any vessel may supplement the whistle signals prescribed in paragraph (*a*) of this Rule by light signals, repeated as appropriate, whilst the manoeuvre is being carried out:

  (i) these light signals shall have the following significance:

    —one flash to mean "I am altering my course to starboard";

    —two flashes to mean "I am altering my course to port";

    —three flashes to mean "I am operating astern propulsion".

  (ii) the duration of each flash shall be about one second, the interval between flashes shall be about one second, and the interval between successive signals shall be not less than ten seconds;

  (iii) the light used for this signal shall, if fitted, be an all-round white light, visible at a minimum range of 5 miles, and shall comply with the provisions of Annex I to these Regulations.

(*c*) When in sight of one another in a narrow channel or fairway:

  (i) a vessel intending to overtake another shall in compliance with Rule 9 (*e*) (i) indicate her intention by the following signals on her whistle:

    —two prolonged blasts followed by one short blast to mean "I intend to overtake you on your starboard side";

    —two prolonged blasts followed by two short blasts to mean "I intend to overtake you on your port side".

  (ii) the vessel about to be overtaken when acting in accordance with Rule 9 (*e*) (i) shall indicate her agreement by the following signal on her whistle:

    —one prolonged, one short, one prolonged and one short blast, in that order.

(*d*) When vessels in sight of one another are approaching each other and from any cause either vessel fails to understand the intentions or actions of the other, or is in doubt whether sufficient action is being taken by the other to avoid collision, the vessel in doubt shall immediately indicate such doubt by giving at least five short and rapid blasts on the whistle. Such signal may be supplemented by a light signal of at least five short and rapid flashes.

(*e*) A vessel nearing a bend or an area of a channel or fairway where other vessels may be obscured by an intervening obstruction shall sound one prolonged blast. Such signal shall be answered with a prolonged blast by any approaching vessel that may be within hearing around the bend or behind the intervening obstruction.

(*f*) If whistles are fitted on a vessel at a distance apart of more than 100 metres, one whistle only shall be used for giving manoeuvring and warning signals.

RULE 35

Sound signals in restricted visibility

In or near an area of restricted visibility, whether by day or night, the signals prescribed in this Rule shall be used as follows:

(*a*) A power-driven vessel making way through the water shall sound at intervals of not more than 2 minutes one prolonged blast.

(*b*) A power-driven vessel underway but stopped and making no way through the water shall sound at intervals of not more than 2 minutes two prolonged blasts in succession with an interval of about 2 seconds between them.

(c) A vessel not under command, a vessel restricted in her ability to manoeuvre, a vessel constrained by her draught, a sailing vessel, a vessel engaged in fishing and a vessel engaged in towing or pushing another vessel shall, instead of the signals prescribed in paragraphs (a) or (b) of this Rule, sound at intervals of not more than 2 minutes three blasts in succession, namely one prolonged followed by two short blasts.

(d) A vessel engaged in fishing, when at anchor, and a vessel restricted in her ability to manoeuvre when carrying out her work at anchor, shall instead of the signals prescribed in paragraph (g) of this Rule, sound the signal prescribed in paragraph (c) of this Rule.

(e) A vessel towed or if more than one vessel is towed the last vessel of the tow, if manned, shall at intervals of not more than 2 minutes sound four blasts in succession, namely one prolonged followed by three short blasts. When practicable, this signal shall be made immediately after the signal made by the towing vessel.

(f) When a pushing vessel and a vessel being pushed ahead are rigidly connected in a composite unit they shall be regarded as a power-driven vessel and shall give the signals prescribed in paragraphs (a) or (b) of this Rule.

(g) A vessel at anchor shall at intervals of not more than one minute ring the bell rapidly for about 5 seconds. In a vessel of 100 metres or more in length the bell shall be sounded in the forepart of the vessel and immediately after the ringing of the bell the gong shall be sounded rapidly for about 5 seconds in the after part of the vessel. A vessel at anchor may in addition sound three blasts in succession, namely one short, one prolonged and one short blast, to give warning of her position and of the possibility of collision to an approaching vessel.

(h) A vessel aground shall give the bell signal and if required the gong signal prescribed in paragraph (g) of this Rule and shall, in addition, give three separate and distinct strokes on the bell immediately before and after the rapid ringing of the bell. A vessel aground may in addition sound an appropriate whistle signal.

(i) A vessel of less than 12 metres in length shall not be obliged to give the above-mentioned signals but, if she does not, shall make some other efficient sound signal at intervals of not more than 2 minutes.

(j) A pilot vessel when engaged on pilotage duty may in addition to the signals prescribed in paragraphs (a), (b) or (g) of this Rule sound an identity signal consisting of four short blasts.

RULE 36

Signals to attract attention

If necessary to attract the attention of another vessel any vessel may make light or sound signals that cannot be mistaken for any signal authorized elsewhere in these Rules, or may direct the beam of her searchlight in the direction of the danger, in such a way as not to embarrass any vessel.

Any light to attract the attention of another vessel shall be such that it cannot be mistaken for any aid to navigation. For the purpose of this Rule the use of high intensty intermittent or revolving lights, such as strobe lights, shall be avoided.

RULE 37

Distress signals

When a vessel is in distress and requires assistance she shall use or exhibit the signals prescribed in Annex IV to these Regulations.

PART E—EXEMPTIONS

RULE 38

Exemptions

Any vessel (or class of vessels) provided that she complies with the requirements of the International Regulations for Preventing Collisions at Sea, 1960, the keel of which is laid or which is at a corresponding stage of construction before the entry into force of these Regulations may be exempted from compliance therewith as follows:

(*a*) The installation of lights with ranges prescribed in Rule 22, until four years after the date of entry into force of these Regulations.

(*b*) The installation of lights with colour specifications as prescribed in Section 7 of Annex I to these Regulations, until four years after the date of entry into force of these Regulations.

(*c*) The repositioning of lights as a result of conversion from Imperial to metric units and rounding off measurement figures, permanent exemption.

(*d*) (i) The repositioning of masthead lights on vessels of less than 150 metres in length, resulting from the prescriptions of Section 3 (*a*) of Annex I to these Regulations, permanent exemption.

(ii) The repositioning of masthead lights on vessels of 150 metres or more in length, resulting from the prescriptions of Section 3 (*a*) of Annex I to these Regulations, until nine years after the date of entry into force of these Regulations.

(*e*) The repositioning of masthead lights resulting from the prescriptions of Section 2 (*b*) of Annex I , to these Regulations until nine years after the date of entry into force of these Regulations.

(*f*) The repositioning of sidelights resulting from the prescriptions of Sections 2 (*g*) and 3 (*b*) of Annex I to these Regulations until nine years after the date of entry into force of these Regulations

(*g*) The requirements for sound signal appliances prescribed in Annex III, to these Regulations until nine years after the date of entry into force of these Regulations.

(*h*) The repositioning of all-round lights resulting from the prescription of Section 9 (*b*) of Annex I to these Regulations, permanent exemption.

ANNEX I

Positioning and technical details of lights and shapes

1. *Definition.*

The term "height above the hull" means height above the uppermost continuous deck. This height shall be measured from the position vertically beneath the location of the light.

2. *Vertical positioning and spacing of lights.*

(*a*) On a power-driven vessel of 20 metres or more in length the masthead lights shall be placed as follows:

    (i) the forward masthead light, or if only one masthead light is carried, then that light, at a height above the hull of not less than 6 metres, and, if the breadth of the vessel exceeds 6 metres, then at a height above the hull not less than such breadth, so however that the light need not be placed at a greater height above the hull than 12 metres;

    (ii) when two masthead lights are carried the after one shall be at least 4·5 metres vertically higher than the forward one.

(*b*) The vertical separation of masthead lights of power-driven vessels shall be such that in all normal conditions of trim the after light will be seen over and separate from the forward light at a distance of 1,000 metres from the stem when viewed from sea level.

(*c*) The masthead light of a power-driven vessel of 12 metres but less than 20 metres in length shall be placed at a height above the gunwale of not less than 2·5 metres.

(*d*) A power-driven vessel of less than 12 metres in length may carry the uppermost light at a height of less than 2·5 metres above the gunwale. When however a masthead light is carried in addition to sidelights and a sternlight or the all-round light of Rule 23 (*c*) (i) is carried in addition to sidelights, then such masthead light or all-round light shall be carried at least 1 metre higher than the sidelights.

(*e*) One of the two or three masthead lights prescribed for a power-driven vessel when engaged in towing or pushing another vessel shall be placed in the same position as either the forward masthead light or the after masthead light; provided that, if carried on the aftermast, the lowest after masthead light shall be at least 4·5 metres vertically higher than the forward masthead light.

(*f*) (i) The masthead light or lights prescribed in Rule 23 (*a*) shall be so placed as to be above and clear of all other lights and obstructions except as described in sub-paragraph (ii).

    (ii) When it is impracticable to carry the all-round light prescribed by Rule 27 (*b*) (i) or Rule 28 below the masthead lights, they may be carried above the after masthead light(s) or vertically in between the forward masthead light(s) and after masthead light(s), provided that in the latter case the requirement of Section 3 (*c*) of this Annex shall be complied with.

(*g*) The sidelights of a power-driven vessel shall be placed at a height above the hull not greater than three-quarters of that of the forward masthead light. They shall not be so low as to be interfered with by deck lights.

(*h*) The sidelights, if in a combined lantern and carried on a power-driven vessel of not less than 20 metres in length, shall be placed not less than 1 metre below the masthead light.

(*i*) When the Rules prescribe two or three lights to be carried in a vertical line, they shall be spaced as follows:

    (i) on a vessel of 20 metres in length or more such lights shall be spaced not less than 2 metres apart, and the lowest of these lights shall, except where a towing light is required, be placed at a height of not less than 4 metres above the hull.

(ii) on a vessel of less than 20 metres in length such lights shall be spaced not less than 1 metre apart, and the lowest of these lights shall, except where a towing light is required, be placed at a height of not less than 2 metres above the gunwale.

(iii) when three lights are carried they shall be equally spaced.

(*j*) The lower of the two all-round lights prescribed for a vessel when engaged in fishing shall be at a height above the sidelights not less than twice the distance between the two vertical lights.

(*k*) The forward anchor light prescribed in Rule 30 (*a*) (i), when two are carried, shall not be less than 4·5 metres above the after one. On a vessel of 50 metres or more in length this forward anchor light shall be placed at a height of not less than 6 metres above the hull.

3.   *Horizontal positioning and spacing of lights.*

(*a*) When two masthead lights are prescribed for a power-driven vessel, the horizontal distance between them shall not be less than one-half of the length of the vessel but need not be more than 100 metres. The forward light shall be placed not more than one-quarter of the length of the vessel from the stem.

(*b*) On a power-driven vessel of 20 metres or more in length the sidelights shall not be placed in front of the forward masthead lights. They shall be placed at or near the side of the vessel.

(*c*) When the lights prescribed in Rule 27 (*b*) (i) or Rule 28 are placed vertically between the forward masthead light(s) and the after masthead light(s) these all-round lights shall be placed at a horizontal distance of not less than 2 metres from the fore and aft centreline of the vessel in the athwartship direction.

(*d*) When only one masthead light is prescribed for a power-driven vessel, this light shall be exhibited forward of amidships, except that a vessel of less than 20 metres in length need not exhibit this light forward of amidships but shall exhibit it as far forward as is practicable.

4.   *Details of location of direction-indicating lights for fishing vessels, dredgers and vessels engaged in underwater operations.*

(*a*) The light indicating the direction of the outlying gear from a vessel engaged in fishing as prescribed in Rule 26 (*c*) (ii) shall be placed at a horizontal distance of not less than 2 metres and not more than 6 metres away from the two all-round red and white lights. This light shall be placed not higher than the all-round white light prescribed in Rule 26 (*c*) (i) and not lower than the sidelights.

(*b*) The lights and shapes on a vessel engaged in dredging or underwater operations to indicate the obstructed side and/or the side on which it is safe to pass, as prescribed in Rule 27 (*d*) (i) and (ii), shall be placed at the maximum practical horizontal distance, but in no case less than 2 metres, from the lights or shapes prescribed in Rule 27 (*b*) (i) and (ii). In no case shall the upper of these lights or shapes be at a greater height than the lower of the three lights or shapes prescribed in Rule 27 (*b*) (i) and (ii).

5.   *Screens for sidelights.*

The sidelights of vessels of 20 metres or more in length shall be fitted with inboard screens painted matt black, and meeting the requirements of Section 9 of this Annex. On vessels of less than 20 metres in length the sidelights, if necessary to meet the requirements of Section 9 of this Annex, shall be fitted with inboard matt black screens. With a combined lantern, using a single vertical filament and a very narrow division between the green and red sections, external screens need not be fitted.

6. *Shapes.*

(*a*) Shapes shall be black and of the following sizes:

  (i) a ball shall have a diameter of not less than 0·6 metre;

  (ii) a cone shall have a base diameter of not less than 0·6 metre and a height equal to its diameter;

  (iii) a cylinder shall have a diameter of at least 0·6 metre and a height of twice its diameter;

  (iv) a diamond shape shall consist of two cones as defined in (ii) above having a common base.

(*b*) The vertical distance between shapes shall be at least 1·5 metres.

(*c*) In a vessel of less than 20 metres in length shapes of lesser dimensions but commensurate with the size of the vessel may be used and the distance apart may be correspondingly reduced.

7. *Colour specification of lights.*

The chromaticity of all navigation lights shall conform to the following standards, which lie with the boundaries of the area of the diagram specified for each colour by the International Commission on Illumination (CIE).

The boundaries of the area for each colour are given by indicating the corner co-ordinates, which are as follows:

  (i) *White*

| x | 0·525 | 0·525 | 0·452 | 0·310 | 0·310 | 0·443 |
| y | 0·382 | 0·440 | 0·440 | 0·348 | 0·283 | 0·382 |

  (ii) *Green*

| x | 0·028 | 0·009 | 0·300 | 0·203 |
| y | 0·385 | 0·723 | 0·511 | 0·356 |

  (iii) *Red*

| x | 0·680 | 0·660 | 0·735 | 0·721 |
| y | 0·320 | 0·320 | 0·265 | 0·259 |

  (iv) *Yellow*

| x | 0·612 | 0·618 | 0·575 | 0·575 |
| y | 0·382 | 0·382 | 0·425 | 0·406 |

8. *Intensity of lights.*

(*a*) The minimum luminous intensity of lights shall be calculated by using the formula:

$$I = 3·43 \times 10^6 \times T \times D^2 \times K\text{-}D$$

where I   is luminous intensity in candelas under service conditions,

T   is threshold factor 2×10^{-7} lux,

D   is range of visibility (luminous range) of the light in nautical miles,

K   is atmospheric transmissivity.

For prescribed lights the value of K shall be 0·8, corresponding to a meteorological visbility of approximately 13 nautical miles.

(*b*) A selection of figures derived from the formula is given in the following table:

Range of visibility (luminous range) of light in nautical miles D	Luminous intensity of light in candelas for $K = 0{\cdot}8$ I
1	0·9
2	4·3
3	12
4	27
5	52
6	94

Note: The maximum luminous intensity of navigation lights should be limited to avoid undue glare. This shall not be achieved by a variable control of the luminous intensity.

9. *Horizontal sectors.*

   (*a*) (i) In the forward direction, sidelights as fitted on the vessel shall show the minimum required intensities. The intensities must decrease to reach practical cut-off between 1 degree and 3 degrees outside the prescribed sectors.

      (ii) For sternlights and masthead lights and at 22·5 degrees abaft the beam for sidelights, the minimum required intensities shall be maintained over the arc of the horizon upto 5 degrees within the limits of the sectors prescribed in Rule 21. From 5 degrees within the prescribed sectors the intensity may decrease by 50 per cent up to the prescribed limits, it shall decrease steadily to reach practical cut-off at not more than 5 degrees outside the prescribed sectors.

   (*b*) (i) All-round lights shall be so located as not to be obscured by masts, topmasts or structures within angular sectors of more than 6 degrees, except anchor lights prescribed in Rule 30, which need not be placed at an impractical height above the hull.

   (*b*) (ii) If it is impracticable to comply with paragraph (*b*) (i) of this section by exhibiting only one all-round light, two all-round lights shall be used suitably positioned or screened so that they appear, as far as practicable, as one light at a distance of one mile.

10. *Vertical sectors.*

   (*a*) The vertical sectors of electric lights as fitted, with the exception of lights on sailing vessels underway shall ensure that:

     (i) at least the required minimum intensity is maintained at all angles from 5 degrees above to 5 degrees below the horizontal;

     (ii) at least 60 per cent of the required minimum intensity is maintained from 7·5 degrees above to 7·5 degrees below the horizontal;

   (*b*) In the case of sailing vessels underway the vertical sectors of electric lights as fitted shall ensure that:

     (i) at least the required minimum intensity is maintained at all angles from 5 degrees above to 5 degrees below the horizontal;

     (ii) at least 50 per cent of the required minimum intensity is maintained from 25 degrees above to 25 degrees below the horizontal.

(*c*) In the case of lights other than electric these specifications shall be met as closely as possible.

11. *Intensity of non-electric lights.*

Non-electric lights shall so far as practicable comply with the minimum intensities, as specified in the Table given in Section 8 of this Annex.

12. *Manoeuvring light.*

Notwithstanding the provisions of paragraph 2 (*f*) of this Annex the manoeuvring light described in Rule 34 (*b*) shall be placed in the same fore and aft vertical plane as the masthead light or lights and, where practicable, at a minimum height of 2 metres vertically above the forward masthead light, provided that it shall be carried not less than 2 metres vertically above or below the after masthead light. On a vessel where only one masthead light is carried, the manoeuvring light, if fitted, shall be carried where it can best be seen, not less than 2 metres vertically apart from the masthead light.

13. *High speed craft.*

The masthead light of high speed craft with a length to breadth ratio of less than 3.0 may be placed at a height related to the breadth of the craft lower than that prescribed in paragraph 2(*a*) (i) of this Annex, provided that the base angle of the isosceles triangles formed by the sidelights and masthead light when seen in end elevation is not less than 27°.

14. *Approval.*

The construction of lights and shapes and the installation of lights on board the vessel shall be to the satisfaction of the appropriate authority of the State whose flag the vessel is entitled to fly.

ANNEX II

Additional signals for fishing vessels fishing in close proximity

1. *General.*

The lights mentioned herein shall, if exhibited in pursuance of Rule 26 (*d*) be placed where they can best be seen. They shall be at least 0·9 metre apart but at a lower level than lights prescribed in Rule 26 (*b*) (i) and (*c*) (i). The lights shall be visible all round the horizon at a distance of at least 1 mile but at a lesser distance than the lights prescribed by these Rules for fishing vessels.

2. *Signals for trawlers.*

(*a*) Vessels of 20 metres or more in length when engaged in trawling, whether using demersal or pelagic gear, shall exhibit:
  (i) when shooting their nets:
      two white lights in a vertical line;
  (ii) when hauling their nets:
      one white light over one red light in a vertical line;
  (iii) when the net has come fast upon an obstruction:
      two red lights in a vertical line.

(*b*) Each vessel of 20 metres or more in length engaged in pair trawling shall exhibit:

    (i) by night, a searchlight directed forward and in the direction of the other vessel of the pair;

    (ii) when shooting or hauling their nets or when their nets have come fast upon an obstruction, the lights prescribed in 2 (*a*) above.

(*c*) A vessel of less than 20 metres in length engaged in trawling, whether using demersal or pelagic gear or engaged in pair trawling, may exhibit the lights prescribed in paragraphs (*a*) or (*b*) of this section as appropriate.

3.   *Signals for purse seiners.*

Vessels engaged in fishing with purse seine gear may exhibit two yellow lights in a vertical line. These lights shall flash alternately every second and with equal light and occultation duration. These lights may be exhibited only when the vessel is hampered by its fishing gear.

ANNEX III

Technical details of sound signal appliances

1.   *Whistles.*

    (*a*) *Frequencies and range of audibility.*

The fundamental frequency of the signal shall lie within the range 70-700Hz.

The range of audibility of the signal from a whistle shall be determined by those frequencies, which may include the fundamental and/or one or more higher freqencies, which lie within the range 180-700 Hz ($\otimes\pm$ 1 per cent) and which provide the sound pressure levels specified in paragraph 1 (*c*) below.

    (*b*) *Limits of fundamental frequencies.*

To ensure a wide variety of whistle characteristics, the fundamental frequency of a whistle shall be between the following limits:

    (i) 70-200 Hz, for a vessel 200 metres or more in length;

    (ii) 130-350 Hz, for a vessel 75 metres but less than 200 metres in length;

    (iii) 250-700 Hz, for a vessel less than 75 metres in length.

    (*c*) *Sound signal intensity and range of audibility.*

A whistle fitted in a vessel shall provide, in the direction of maximum intensity of the whistle and at a distance of 1 metre from it, a sound pressure level in at least one 1/3rd-octave band within the range of frequencies 180-700 Hz ($\pm$ 1 per cent) of not less than the appropriate figure given in the table below.

1/3rd octave band level at 1 metre in Length of vessel in metres	Audibility dB referred to 2×10^{-5} N/m^2	range in nautical miles
200 or more	143	2
75 but less than 200	138	1·5
20 but less than 75	130	1
Less than 20	120	0·5

The range of audibility in the table above is for information and is approximately the range at which a whistle may be heard on its forward axis with 90 per cent probability in conditions of still air on board a vessel having average background noise level at the listening posts (taken to be 68 dB in the octave band centred on 250 Hz and 63 dB in the octave band centred on 500 HZ).

In practice the range in which a whistle may be heard is extremely variable and depends critically on weather conditions; the values given can be regarded as typical but under conditions of strong wind or high ambient noise level at the listening post the range may be much reduced.

(*d*) *Directional properties.*

The sound pressure level of a directional whistle shall be not more than 4 dB below the prescribed sound pressure level on the axis at any direction in the horizontal plane within ± 45 degrees of the axis. The sound pressure level at any other direction in the horizontal plane shall be not more than 10 dB below the prescribed sound pressure level on the axis, so that the range in any direction will be at least half the range on the forward axis. The sound pressure level shall be measured in that 1/3rd-octave band which determines the audibility range.

(*e*) *Positioning of whistles.*

When a directional whistle is to be used as the only whistle on a vessel, it shall be installed with its maximum intensity directed straight ahead.

A whistle shall be placed as high as practicable on a vessel, in order to reduce interception of the emitted sound by obstructions and also to minimize hearing damage risk to personnel. The sound pressure level of the vessel's own signal at listening posts shall not exceed 110 dB (A) and so far as practicable should not exceed 100 dB (A).

(*f*) *Fitting of more than one whistle.*

If whistles are fitted at a distance apart of more than 100 metres, it shall be so arranged that they are not sounded simultaneously.

(*g*) *Combined whistle systems.*

If due to the presence of obstructions the sound field of a single whistle or of one of the whistles referred to in paragraph 1 (*f*) above is likely to have a zone of greatly reduced signal level, it is recommended that a combined whistle system be fitted so as to overcome this reduction. For the purposes of the Rules a combined whistle system is to be regarded as a single whistle. The whistles of a combined system shall be located at a distance apart of not more than 100 metres and arranged to be sounded simultaneously. The frequency of any one whistle shall differ from those of the others by at least 10 Hz.

2.   *Bell or gong.*

(*a*) *Intensity of signal.*

A bell or gong, or other device having similar sound characteristics shall produce a sound pressure level of not less than 110 dB at a distance of 1 metre from it.

(*b*) *Construction.*

Bells and gongs shall be made of corrosion-resistant material and designed to give a clear tone. The diameter of the mouth of the bell shall be not less than 300 mm for vessels of 20 metres or more in length, and shall be not less than 200 mm for vessels of 12 metres or more but of less than 20 metres in length. Where practicable, a power-driven bell striker is recommended to ensure constant force but manual operation shall be possible. The mass of the striker shall be not less than 3 per cent of the mass of the bell.

3. *Approval.*

The construction of sound signal appliances, their performance and their installation on board the vessel shall be to the satisfaction of the appropriate authority of the State whose flag the vessel is entitled to fly.

ANNEX IV

Distress signals

1. The following signals, used or exhibited either together or separately, indicate distress and need of assistance:

(a) a gun or other explosive signal fired at intervals of about a minute;

(b) a continuous sounding with any fog-signalling apparatus;

(c) rockets or shells, throwing red stars fired one at a time at short intervals;

(d) a signal made up by radiotelegraphy or by any other signalling method consisting of the group . . . — — — . . . (SOS) in the Morse Code;

(e) a signal sent by radiotelephony consisting of the spoken word "Mayday";

(f) the International Code Signal of distress indicated by N.C.;

(g) a signal consisting of a square flag having above or below it a ball or anything resembling a ball;

(h) flames on the vessel (as from a burning tar barrel, oil barrel, etc.);

(i) a rocket parachute flare or a hand flare showing a red light;

(j) a smoke signal giving off orange-coloured smoke;

(k) slowly and repeatedly raising and lowering arms outstretched to each side;

(l) the radiotelegraph alarm signal;

(m) the radiotelephone alarm signal;

(n) signals transmitted by emergency position-indicating radio beacons.

(o) approved signals transmitted by radiocommunication systems, including survival craft radar transponders.

2. The use or exhibition of any of the foregoing signals except for the purpose of indicating distress and need of assistance and the use of other signals which may be confused with any of the above signals is prohibited.

3. Attention is drawn to the relevant sections of the International Code of Signals, the Merchant Ship Search and Rescue Manual and the following signals:

(a) a piece of orange coloured canvas with either a black square and circle or other appropriate symbol (for identification from the air);

(b) a dye marker.

S

"S is for Stunsails, aloft and alow".

Sagged — A vessel is sagged when she is strained and her midships part is lower than her bow and stern.

Salute — The naval salute is made by bringing the right hand up to the cap, thumb and fingers together and in line with hand and forearm, palm of the hand turned downwards, elbow level with the shoulder.

Sails and Sailing — The remarks that follow refer chiefly to ship's lifeboats. These boats, being built to be stable and seaworthy when full of people, are not handy or weatherly craft. Their sail area is small (mast and sails must be stowed in the boat) and their freeboard is high and rudder not deeply immersed when not loaded. A yacht will certainly handle more easily and certainly sail much closer to the wind. On the other hand a yacht may need reefing earlier because of her initially larger sail area.

Rigs — Most ship's boats were sloop rigged with a mainsail and foresail (or jib). (*See* Figs. 1 and 2).

Some vessels are rigged with two masts, the taller always being known as the mainmast. When the shorter mast is abaft the main (as in the ketch and yawl rig) it is called the **mizzen** and the sail set upon it the mizzen sail. (*See* Fig. 3).

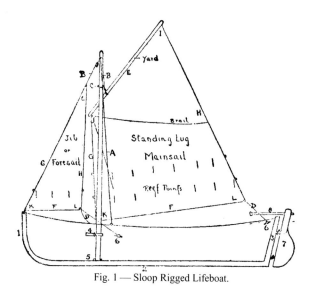

Fig. 1 — Sloop Rigged Lifeboat.

Parts of the boat	Rigging	Parts of the sails
1. Stem.	*A* Shroud.	*E* Head, the upper edge.
2. Keel.	*B* Halyards.	*F* Foot, the lower edge.
3. Sternpost.	*C* Traveller.	*G* Luff, the foremost edge.
4. Mast thwart.	*D* Sheets.	*H* Leech, the after edge.
5. Mast step.		*I* Peak, upper after corner.
6. Cleats.		*J* Throat, upper foremost
7. Rudder (note pintles		corner.
and gudgeons).		*K* Tack, lower foremost
8. Tiller.		corner.
		L Clew, lower after corner.

Reef points are pieces of the line worked through the sail for shortening sail.

Cringles (holes in the bolt-rope fitted with thimbles are worked into each corner of the sail and at the end of each row of reef points.

The curvature of the foot of a sail is known as the roach.

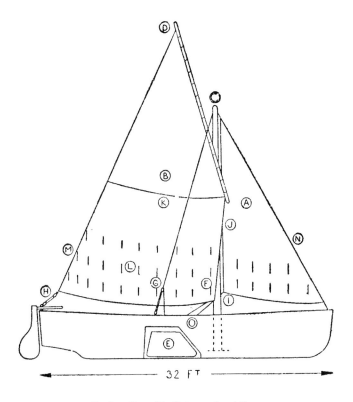

Fig. 2 — Sloop Rig Cutter — Royal Navy.

A	Foresail.	I	Clew.
B	Mainsail.	J	Halyards (leading down alongside
C	Mast.		Mast).
D	Yard.	K	Brail.
E	Drop Keel.	L	Reef Points.
F	Shroud (not shown. Supports	M	Cringles.
	mast athwartships).	N	Forestay.
G	Runner.	O	Fore Sheets.
H	Main Sheet.		

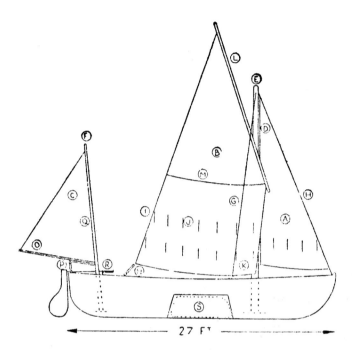

Fig. 3 — Whaler.
ROYAL NAVY — There is now a multi-purpose whaler fitted with an engine.

A	Foresail.	J	Reef Points.
B	Mainsail.	K	Fore Sheets.
C	Mizzen.	L	Yard.
D	Halyards (leading down alongside Mast).	M	Brail.
		N	Main Sheet.
E	Mainmast.	O	Mizzen Boom.
F	Mizzenmast.	P	Mizzen Sheets.
G	Shroud.	R	Sailing Tiller.
H	Forestay.	S	Drop Keel.
I	Cringle.		

Schooners, which have the shorter mast forward, have a **foremast** and a mainmast and set a foresail and mainsail. All, of course, carry headsails: jib, or jibs, and a forestaysail. Schooners may also carry a main topmast staysail. Staysails take their names from the stays on which they are hoisted.

The **Mainsail** of a ship's boat is usually a standing lug. The head of the sail is laced to a yard, and a strop by which it is hoisted is fitted round the yard at about a quarter of its length from forward. The tack is made fast at the foot of the mast, and the sheet, by which the sail is trimmed, is led aft to a cleat.

Ship's boats must have a mast short enough to stow inside them. Yachts are not usually restricted in this way and the mast is often longer than the boat. This enables the sail to have a very long luff and the sail is "jib-headed" or **bermuda** type. The leech of a bermuda mainsail is usually stiffened with battens near the top which ship into pockets made in the sail.

Older yachts set a mainsail on a gaff, which is a spar fitted with jaws to take the mast and hoisted by two halyards at the peak and the throat. Gaff mainsails and sometimes standing lug mainsails are furled by means of brails which are lines seized to the leach of the sail, leading through thimbles near the throat and down into the boat and which haul up the sail like a curtain. The lee brail is always hauled in best to prevent this sail from bagging and holding the wind.

Small boats may be fitted with a **dipping lug** and no jib. The tack of the dipping lug is hooked to a tack hook near the stem (not at the mast like a standing lug) and has to be dipped round the mast to the lee side when the boat goes about.

Gunter Rig — Combines the advantage of a short mast and high-peaked sail. The lower end of the gaff is fitted with jaws and a parrel which goes round the mast and the halyard makes fast to a wire span fitted to the gaff. When hoisted the gaff is almost vertical and projects well above the masthead.

Booms — Bermuda mainsails always have, and other sails may have, a boom to stretch the foot of the sail. The forward end is hinged by a gooseneck to the mast and the after end is supported by a topping lift leading well up the mast and then down on deck. (Compare the rigging of a boom with that of a derrick). The main sheet is attached to the boom.

Reefing the Mainsail — The vessel may first be hove-to to shorten sail, settle (*i.e.* lower slightly), the halyards, make the tack and clew fast to the reef cringles, bunch up the sail, tie the reef points and reset the sail. When shaking out the reef, be sure *all* the reef points are untied before the sail is hoisted or it will be torn.

Roller-reefing gear maybe fitted to yachts. The boom is rotated by shipping a reefing handle or crank near the goose-neck. The halyard is then slowly eased and the handle turned so that the sail is wound round the boom like a blind. The weight of the boom may first be taken by the topping lift which is slackened again when the reefing is completed. Always reef in good time.

The Headsails: Foresail or Jib — The halyard usually leads through a block at the masthead. The tack is made fast to the stem head. The sail is fitted with two sheets, one being led aft on each side of the boat and the weather sheet always being kept slack.

Sails are partly roped round the edges with bolt-rope to strengthen them, and as the roping is always sewn on the port side of the sail this enables one to see that the sail is bent on the right way round.

Jibs are not usually reefed. Yachts may carry several of different sizes, the size suitable for the weather being used. The largest size; No. 1, is called the **genoa** and may be as big as the mainsail. The smallest is called the storm jib. The different jibs are stowed in bags and kept below until needed. When wanted a bag is passed up on deck and made fast (so that it will not blow overboard when emptied). The tack of the sail is first brought out and made fast and then the hanks along the luff to the stay. Finally the halyard and sheet are shackled to the head and clew. When stowing the sail the tack goes into the bag last so that it will be on top when next required.

Boats may have a storm trysail to be used instead of the mainsail in very bad weather. The trysail can also be used in light weather when running before the wind as a spinnaker (*See* next paragraph). When hoisted as a spinnaker the tack is made fast near the foot of the mast and the clew borne out on the opposite side to the mainsail by means of a boathook or bearingout spar, greatly increasing the sail area.

Spinnaker — A light sail of symmetrical triangular shape, used when the wind is abaft the beam and set on the opposite side of the boat to the mainsail, *i.e.* on the windward side of the boat. It is kept carefully stowed in its bag with the three corners on top ready for bending on. The halyard comes down from near the masthead and the sail sets ahead of the forestay. The foot of the sail is extended by the spinnaker boom, one end of which ships on to a fitting on the mast and the other is attached to a topping lift and downhaul. The tack of the sail is attached to the outer end of the boom. Two ropes — the sheet and the guy — lead from aft, outside everything, the sheet on the lee side to the clew of the sail, and the guy to the outer end of the boom.

Masts — The heel of (a lifeboat's) mast fits into a step cut for it in the keelson, and it is secured to the mast thwart by a clamp. Be careful to step the mast the right way round or the halyards will lead foul; most masts are fitted with a tack hook to which the main tack is made fast and this is always on the afterside of the mast, and a white line is painted on the fore side of the mast at the level of the thwart.

When making a long passage before the wind in a lifeboat, the mast should be stepped the other way round *i.e.*, with the tack hook forward, so that the main halyards will lead from forward aft and thus form a backstay.

Rigging — The mast of a lifeboat is strengthened by two wire shrouds which fit over the masthead and are made fast to the gunwale by lanyards. The loftier masts of cutters and larger boats are supported in addition to the shrouds by a **forestay**, leading from the masthead to the stem, and two **backstays** which set up to the gunwale abaft the shrouds. **Halyards**, by which the sails are hoisted, are rove through sheaves or blocks at the masthead. Be careful not to let the end of a halyard go, or it may run up the mast out of reach. One end of the main halyard is spliced to an iron ring which runs up and down the mast, called a **traveller**. The strop on the yard is hooked to the traveller when the sail is hoisted.

Setting Sail (in a lifeboat) — Step the mast and set up the shrouds, make the tack of the mainsail fast at the tack hook on the mast and the tack of the foresail fast to the stemhead. Bend on the halyards and see that everything is clear before hoisting. When leaving the ship's side under sail pass the painter aft on the side nearest the ship so that the boat's head swings well out before letting go.

Coming Alongside — See that the halyards are clear for running. Haul the jib down first, thus ensuring that the boat will come head to wind when the helm is put down.

Sailing Close-hauled and Running Free — A boat cannot, of course, sail directly against the wind. When she is sailing as close to it as she can (which is about 5 or 6 points away, or 4 points for a yacht), she is said to be **close-hauled**. When sailing close-hauled with the wind on the starboard side she is said to be on the starboard tack, and when close-hauled with the wind on the port side, on the port tack. The coxswain steers so that there is a faint quiver in the throat of the sail. If the sail begins to shake it is not drawing properly and the boat is too close to the wind and must be kept away, while if the quiver disappears the boat can be made to come closer to the wind. It does not pay to keep a ship's boat too close to the wind, or "pinch her", as she makes too much leeway (*i.e.*, drift to leeward). It is better to keep her "full and by" the wind when she will sail faster and make less leeway, than to "starve her" and keep her as close to the wind as possible.

When the wind blows from abaft the beam the sheets are eased away. When the wind is nearly aft the coxswain must be careful not to **gybe**. This means that, from careless steering or a shift of wind, the wind gets on the other side of the sail and the sail swings across to the other side of the boat. In a strong breeze this may dismast or capsize the boat.

When sailing first on one tack and then on the other the boat is said to be **beating to windward**. When the wind is about abeam, she is **reaching** or **on a reach**. If she can sail to her destination on one reach and return on the other she is said to have a **soldier's wind**. When sailing with the wind on the quarter she is said to be **running** or **running free.**

"Helm Up" and "Helm Down" — The coxswain always sits on the weather side of the helm or tiller because from that side he gets a better view, and as the boat always lists to leeward he therefore always sits on the higher side of the boat. When he puts the helm "down" he pushes the tiller or helm away from him *down* to the lower side of the boat, and the boat's head swings to windward. When he puts the helm "up" he pulls it *up* towards him and the boat's head swings to leeward. These terms are always used when sailing, except when the wind is right aft when "port" and "starboard" are used indicating which way the boat's head should move. (*See* Steering).

Weather Helm and Lee Helm — When a boat requires the helm to be kept a little "a-weather" or "up" to keep her on a straight course, she is said to carry weather helm, and if it has to be kept to leeward or "down", lee helm. A boat should carry a little weather helm to ensure that she will come up in the wind when required. If she does not, move the crew further forward in the boat.

To luff means to put the helm down and bring the boat's head towards the wind.

To keep away or **bear away** means to put the helm up and turn the boat's head away from the wind, to leeward.

Tacking — When it is necessary to go in the direction from which the wind is blowing, the boat has to steer in a series of zig-zags or tacks close-hauled with the wind first on one side and then on the other. The process of getting from one tack to the other is called **going about** (Fig. 4). Before going about the boat is kept a little off the wind to get plenty of way on. Then the cautionary order "ready about" is given and the coxswain puts the helm down (do not jam the helm over suddenly, push it down steadily), and the main sheet is hauled in, and the boat swings head to wind (if the boat is sluggish and does not come head to wind easily the jib sheet should be eased at the same time). When in this position the jib becomes aback (*i.e.*, with the wind on its fore side) and helps the boat's head round still further. As soon as the mainsail fills on the new tack the order "let draw" or "jib sheet over" is given and the jib sheeted home on the lee side. It often happens that a lifeboat will not sail round on to the other tack which is called **missing stays**, or hangs head to wind refusing to pay off on either tack, known as being **in irons**. In these cases it is usual to ship an oar over the lee side and pull her head round.

Wearing — If a boat refuses to go about head to sea she must be brought from one back to the other by "wearing" or going round stern to wind. This is not otherwise usually done as it loses a lot of ground, and, in a strong breeze, gybing is dangerous. To wear, the helm is put up and the boat run off before the wind. As the wind comes aft the main sail is gybed,

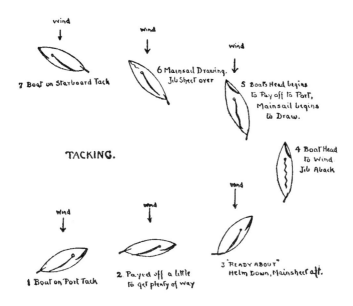

wind

wind

wind

6 Mainsail Drawing.
Jib Sheet over

5 Boats Head begins
to Pay off to Port,
Mainsail begins
to Draw.

7 Boat on Starboard Tack

4 Boat Head
to Wind
Jib Aback

TACKING.

wind

wind

wind

3 "READY ABOUT"
Helm Down, Mainsheet aft.

1 Boat on Port Tack

2. Payed off a little
to get plenty of way

Fig. 4 — Tacking.

care being taken to haul in the mainsheet and pay it out to prevent the sail taking charge and going across with a jerk. In a strong breeze the sail should be brailed up (if brails are fitted) or the sail lowered right down and rehoisted on the opposite side. The helm is kept over and the boat brought close to the wind again.

Squalls — When sailing close-hauled, luff the boat slightly, easing the sheet in order to spill some of the wind if necessary. When running free keep the boat more before the wind, when she will sail faster and heel over less. When running before a heavy sea, run the boat off at each wave so that it strikes the stern; if it catches the quarter the boat may broach to (*i.e.*, get thrown broadside on to the sea) and capsize.

With a strong wind nearly aft the mainsail may be hauled down, the boat continuing to run safely under the jib alone. If the weather continues to worsen, the boat should be hove to, *i.e.*, kept stationary, or almost stationary with the sea-anchor over the stern (in a double-ended boat) and the oil bag used. The mast (and jib, if set or partly set) will help to keep her stern to sea.

Never hitch the sheets. In a small boat take a turn round a cleat and hold them. If hitched they are sure to jam when you want to ease them in a gust or squall, and the boat may thus be capsized. Never let them go altogether.

Always have the halyards and sheets coiled down clear for running, gear "squared up" neatly and fenders inboard.

Never walk about on the thwarts. Remember that the crew forms the ballast and that you must keep the boat properly trimmed. The proper place for the crew when sailing is in the bottom of the boat abaft the mast. (*See* Boats and Boatwork Handling Craft; Lifesaving Appliances).

Sailmaking — To join two cloths together, lay the selvage of one cloth along the blue line worked in the canvas $1\frac{1}{2}$ inch from the selvage of the other, and, if the seam is a long one, tack it together every few feet. Secure the right end of the canvas with the hook. A *Flat Seam* is made working from right to left, pushing the needle through the cloth towards you. To sew a *Round Seam* first rub down the canvas at the blue line, and working from left to right, sew the cloth to the bight you have rubbed down pushing the needle away from you. To mend a tear, first join the torn edges with a herring-bone stitch and then patch it, using a flat seam.

Samson Post — A stump mast used to support a derrick.

Scotchman — A piece of steel, wood or hide fitted to prevent chafe.

Scouts, Deep Sea — Scouts who are members of the Royal Navy, Merchant Navy, Fishing Fleets, or of Sea Training Establishments.

Scud — To run rapidly before the wind; thin clouds driven before the wind.

Scuppers — Holes or pipes in a vessel's sides for draining the water from the decks.

Scuttles — Portholes.

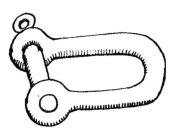

Straight Screw Shackle.

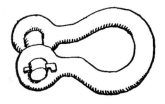

Bow Shackle with Forelock.

Seafarers Education Service — Provides libraries to ships, and tuition books and help with studies and hobbies to all seafarers. Address: 202 Lambeth Road, London SE1 7LQ.

Sennit — Plaited rope.

Settle — To lower a little.

Sextant — An instrument for measuring angles, used in fixing a ship's position.

Shackle — The part in which the pin fits is known as the lug of the shackle. A cable's joining shackles have their bolts or pins secured by a wooden peg or steel pin passing through lug and bolt. Other shackles have screwed pins, while some are fitted with forelocks which fit in to the end of the pins.

Sheer — The curve upwards of a ship's upper deck towards the bow and stern. The movement of a vessel to port or starboard when at anchor.

Shifting Boards — Temporary wooden centre-line bulkheads rigged in the 'tween decks and holds before a bulk cargo of grain is loaded to prevent the grain shifting in bad weather and giving the vessel a list.

Shipshape — Trim, in a seamanlike manner.

Ships — A seaman should be able to recognise the different types of ships at sight.

WARSHIPS
The main types are:

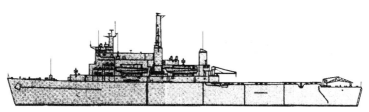

Amphibious Ship

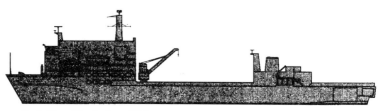

Aviation Training Ship

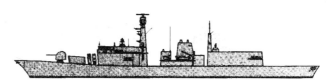

Type 23 Frigate

Type 22 Frigates   Batch 3

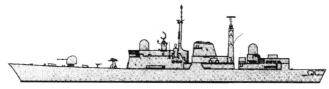

Type 42 Destroyers   Batch 3 (Stretched)

Fig. 5

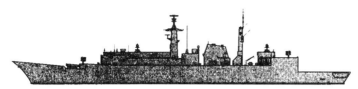

Type 22 Frigates   Batch 2

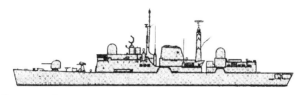

Type 42 Destroyers   Batch 1 and 2

Fleet Submarines

Trafalgar Class

Polaris Submarines

Trident Class

Fig. 6

Aircraft Carriers — Are easily distinguished by their long flat flight decks, hence their nickname of "flat-tops". Displacement 20,000 to 50,000 tons. Length 700 to 1000 feet.

Cruisers — General-purpose fighting ships. Displacement 12,000 to 14,000 tons; length 550 to 700 feet. The new thorough-deck command cruisers will carry helicopters and perhaps vertical-take-off aircraft, so they will have a conspicuous flight-deck. They will be powered by gas-turbines and armed with missiles.

Destroyers — Guided missile destroyers, displacement over 5000 tons; length about 550 feet; steam and gas turbines; armed with missiles, guns, torpedoes and a helicopter. (Figs. 5 and 6).

Frigates — Displacement about 2000 tons; length about 360 feet; steam and gas turbines. There are four main types: General-purpose frigates for escort work, giving anti-submarine and anti-aircraft protection; the Anti-submarine (A/S) frigate; the Anti-aircraft (A/A) frigate and the Air Defence (A.D.) frigate for the direction of aircraft. (Figs. 5 and 6).

Submarines — Displacement 2000 tons and upwards; length 250 to 300 feet. Nuclear-powered turbines or diesel and electric engines. Nuclear-powered submarines can spend very long periods completely submerged and are of high speed. The nuclear submarines fitted with Trident and Polaris missiles can hit a target 2500 miles away. Fleet submarines are armed principally with torpedoes. (Fig. 6).

Commando and **Assault ships** carry military forces; **Mine Counter measures vessels** (Coastal Mine-sweepers and Coastal Minehunters); **Fleet Support** ships, and **Royal Fleet Auxiliaries** (R.F.A.) etc.

MERCHANT SHIPS
Passenger Ships — Can easily be distinguished from cargo ships by their additional superstructure and greater number of boats. (Fig. 7).

Cargo Ships — A typical cargo steamer is shown in Fig. 8).

Fig. 7 — RMS *Windsor Castle*.

Fig. 8 — A Refrigerated Cargo Ship.

Fig. 9 — Yawl.

Container Ship — A ship constructed with slides in its holds to facilitate the loading and discharge of containers. When the hatches are closed, containers are then stacked four or even five high on deck. Because the early containers were mostly twenty feet long the capacity of a container ship may be given in twenty foot equivalent unit T.E.U.S.

Motor Ships — Can often be recognised by their squat funnels.

Tankers — Ships carrying oil in bulk; are distinguished by having the funnel aft and the absence of many derricks. Some container ships and coasters also have their funnel aft.

FISHING CRAFT

Trawlers — Are built with a bold sheer; and have their funnel abaft amidships and a prominent wheelhouse in front of it. Stern trawlers have a conspicuous gantry aft for handling the trawl.

Drifters — Are smaller than trawlers and, when lying to their nets, often lower their foremasts.

SAILING CRAFT

Ships — Are vessels with three masts, square-rigged on all three masts. The word ship is nowadays used to denote any large vessel, however propelled.

Barques — Are three-masted vessels square-rigged on the fore and main masts and fore-and-aft rigged on the mizzen mast. Four-masted barques are square-rigged on the fore, main and mizzen masts and fore-and-aft rigged on the fourth mast which is called the jigger mast. (Fig. 10).

Barquentines — Three-masted vessels square-rigged on the foremast. (Fig. 12).

Brigs — Two-masted vessels, square-rigged on both masts.

Brigantines — Two-masted vessels, square-rigged on the foremast.

Schooners — Two or more masted fore-and-aft rigged vessels. (Fig. 13).

Topsail Schooners — The same schooners, but carry square topsails on the foremast. (Fig. 11).

Fig. 10 — Steel Four Mast Barque.

Fig. 11 — Top's'l Schooner *Sir Winston Churchill*.

Fig. 12 — Barquentine *Waterwitch*.

Fig. 13 — Grand Banks Schooner *Bluenose*.

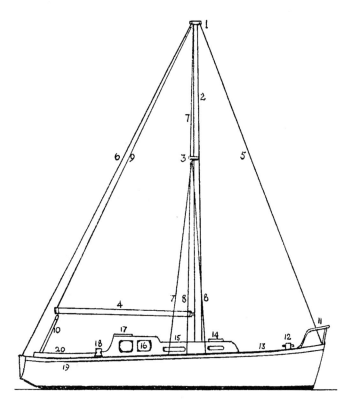

YACHT.

1.	Truck.	11.	Pulpit.
2.	Mast.	12.	Samson post.
3.	Spreader.	13.	Fore deck.
4.	Boom.	14.	Fore hatch.
5.	Forestay.	15.	Coachroof.
6.	Backstay.	16.	Doghouse.
7.	Topmast shroud.	17.	Main hatch.
8.	Lower shrouds.	18.	Sheet winch.
9.	Topping lift.	19.	Cockpit.
10.	Main sheet.	20.	Cockpit coaming.

Fig. 14.

Cutters — One-masted fore-and-aft rigged vessels.

Sloops — The same as cutters, but have no bowsprit (or only a very short one) and only one headsail (*e.g.*, ships' boats). (Figs. 1 and 2).

Yawls — The same as cutters but have also a small mizzen mast stepped right aft.

Ketches — Like yawls, but the mizzen is larger and stepped forward of the rudder.

Paintwork — Warships are painted grey. Liners engaged in cruising and yachts are usually painted white. Merchant ships generally are black with white upper-works. Ships of different lines can be distinguished by the colours of their funnels. For example:

Black funnels — P&O (those ships with white hulls have yellow funnels). *Black funnels with white bands* — British India S.N.C. *Black funnel with 2 red bands* — Clan Line.

Yellow funnels — Orient, Canadian Pacific, Royal Mail Lines. New Zealand Shipping Co.

Yellow funnels with black tops — Shaw, Savill & Albion.

Yellow funnel with white band and black top — Ellerman Lines.

Red funnels with black tops — Cunard, Union Castle, Port Line, etc.

Ship Construction — In order to make ships as unsinkable as possible, the hull (the main body of the ship) is divided into watertight compartments by partitions known as **Bulkheads**. These bulkheads are placed between each hold, at each end of the engine and boiler rooms, and right forward as a collision bulkhead. For the same reason ships are built with **Double-Bottoms**, the compartments between the inner and outer bottom being used to store oil fuel, fresh water or water ballast. Older oil tankers although divided into possibly 27 tanks still represent a hazard in the event of a collision or stranding. As they have a single hull, any damage leads to an oil spill with consequent risk to the marine environment and wildlife. The disposal of oily ballast water is another problem and the designers of the latest tankers combat these two difficulties by having double bottoms and constructing segregated ballast tanks at the sides of the ship. In this way the ballast water stays clean and if the outer hull is pierced the inner hull should still contain the oil cargo.

The steel plating of a ship is built over, and strengthened by, a framework or skeleton of steel which is arranged in much the same way as in a boat (compare it with a boat's construction on page 21 and Fig. 16). The **keel** is (except in small vessels) a flat plate upon which stands the vertical plate **Keelson**.

Pages 166-172 show a selection of ships with differing styles and uses. ▶

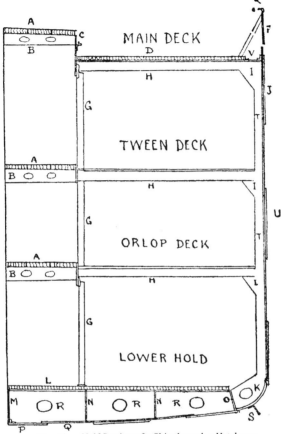

Fig. 16 — Half Section of a Ship through a Hatchway.

A	Wooden hatches.
B	Movable hatch beam (holes cut for lightening purposes).
C	Hatch coaming.
D	Wooden deck laid over steel deck.
E	Main rail.
F	Bulwarks.
G	Stanchions or pillars.
H	Beams.
I	Beam knees joining beam to frame.
J	Sheer strake.
K	Bracket in bilge joining frame to margin plate.
L	Wooden ceiling laid over tank top.
M	Keelson.
N	Side keelsons.
O	Margin plate, separating double bottom from bilge.
P	Keel.
Q	Garboard strake.
R	Floor, dividing the double bottom athwartships.
S	Bilge keel.
T	Frame.
U	Plating.
V	Waterway.

Running athwartships are vertical plates called **Floors** and between these, and parallel to the keelson, are **Side Keelsons**. The inner and outer bottoms are riveted to this framework. The wing plates of the inner bottom are known as **Margin Plates** and often slope down to the ships' side forming the **Bilges** to which any water in the hold is allowed to drain. In many ships the margin plates do not slope downwards and the water drains to wells in the double bottom.

Joining the margin plate by means of large brackets are the **Frames** to which the side plating is attached. The frames are tied together in the fore-and-aft direction by long **Stringers** and at the top the frames are connected by **Beams** which carry the decks. The decks are further supported and tied together by **Pillars** or **Stanchions**.

The **Stempost** is scarphed into the fore end of the keel, and the after end of the keel is riveted to the **Stern Frame**, a massive casting which supports the rudder and propeller.

To this framework the plating is welded or riveted in long fore-and-aft strakes. Welding is now the usual method of joining steel plates and the strakes are laid flush to present a smooth surface. If they are riveted together they overlap, one strake overlapping or being overlapped by two others. At the turn of the bilge, **Bilge Keels** are fitted which help to steady the ship without increasing her draught.

For Glass Reinforced Plastic and Wood Construction (*See* under Boats and Boatwork).

SIGNALS AND SIGNALLING

INTERNATIONAL CODE OF SIGNALS, 1969

The *International Code* enables ships to communicate with each other no matter what their nationality and language may be.

The signals used consist of:
   (a) Single-letter signals allocated to significations which are very urgent, important, or of very common use;
   (b) Two-letter signals for the General Service;
   (c) Three-letter signals beginning with "M" for the Medical Section. These are printed upon green paper.

The International Code of Signals is augmented from time to time by Amendments. Amendment List No. 2, for instance, contained, among other additional signals, the following:
   YG   You appear not to be complying with the traffic separation scheme.

Each signal has a complete meaning. **Complements** in certain cases are used to supplement the available groups. The complements vary the meaning of the basic signal.

INTERNATIONAL CODE OF SIGNALS

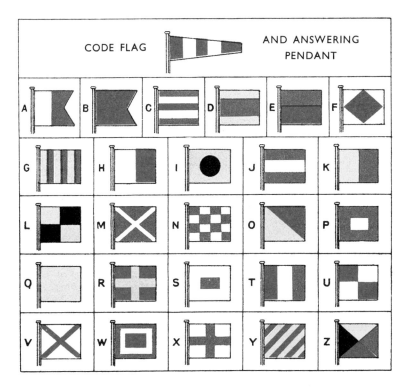

NUMERAL PENDANTS.

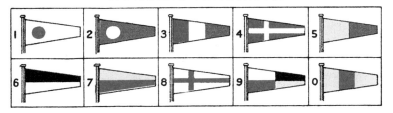

SUBSTITUTES.

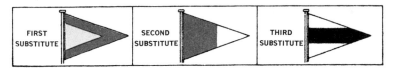

NATIONAL COLOURS

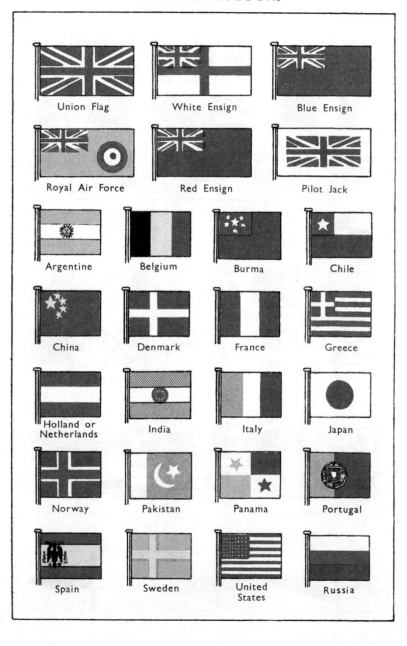

Union Flag   White Ensign   Blue Ensign

Royal Air Force   Red Ensign   Pilot Jack

Argentine   Belgium   Burma   Chile

China   Denmark   France   Greece

Holland or Netherlands   India   Italy   Japan

Norway   Pakistan   Panama   Portugal

Spain   Sweden   United States   Russia

Station in the code means a ship, aircraft, survival craft or place at which communications can be effected.

Time of Origin is the time at which a signal is ordered to be made. If made at the end of the text it serves as a convenient reference number.

Identity Signal or call sign is the group of letters and figures assigned to each station by its administration. British ships' signal letters begin with either *G* or *M* and are printed in *Signal Letters of British Ships*. Example: *GJRN* is the Identification Signal and Radio Call Sign of the *British Captain*.

SINGLE-LETTER SIGNALS

May be made by any method of signalling. Those marked * when made by sound may only be made in compliance with the requirements of the *International Regulations for Preventing Collisions at Sea*, Rules 34 and 35. (The notes in brackets are intended to serve as aids to memory).

A   I have a diver down; keep well clear at slow speed. (*A*qualung).

B   I am taking in, or discharging, or carrying dangerous goods. (*B*eware or you may be *B*lown up. Bang!)

C   Yes: affirmative; or "The significance of the previous group should be read in the affirmative". (*C*ertainly. Si (Spanish) = "Yes").

*D**   Keep clear of me, I am manoeuvring with *D*ifficulty.

*E**   I am altering my course to starboard. (One short blast).

F   I am disabled; communicate with me. (Engine *F*ailure).

G   I require a pilot. (*G*ive me a pilot). When made by fishing vessels operating in close proximity on the fishing grounds it means: "I am hauling nets".

*H**   I have a pilot on board.

*I**   I am altering course to port. (Two short blasts).

J   I am on fire and have a dangerous cargo on board: keep well clear of me.

K   I wish to communicate with you. (*K*ommunication).

L   You should stop your vessel instantly.

M   My vessel is stopped and making no way through the water.

N   No: negative; or "The significance of the previous group should be read in the negative". This signal may be given visually or by sound. For voice or radio transmission the signal should be "NO".

O   Man *O*verboard.

P   In harbour: All persons should report on board as the vessel is about to proceed to sea. (Blue *P*eter).
    At Sea: It may be used by fishing vessels to mean "My nets have become fast upon an obstruction".

Q   My vessel is "healthy" and I request free pratique.

R*   I am at anchor; you may feel your way past me.†
S*   My engines are going astern. (3 short blasts).
T*   Keep clear of me; I am engaged in pair trawling. (*T*rawlers).
U   You (*U*) are running into danger.
V   I require assistance. *V* comes after *U*!)
W   I require medical assistance. (*W*ounds).
X   Stop carrying out your intentions and watch for my signals.
Y   I am dragging my anchor.
Z   I require a tug. When made by fishing vessels operating in close proximity on the fishing grounds it means "I am shooting nets".

† This signal is not in the *International Code*, but in Rule 35 of the *International Regulations for Preventing Collisions*.

Single-letter signals are very often used; the most common are:

Flag Signals
B   Dangerous goods on board.
G   "I require a pilot".
H   "I have a pilot on board". A white and red flag divided horizontally indicates that a pilot licenced by Trinity House is on board.
P   The Blue Peter. "I am about to proceed to sea".
Q   "My vessel is healthy and I request free pratique".
Flown by a vessel newly arrived from another country.

Sound Signals — (derived from *The Regulations for Preventing Collisions at sea).*
(i) *Manoeuvring Signals*
E   "I am altering course to starboard".
I   "I am altering course to port".
S   "My engines are going astern".
D   "Keep clear of me: I am manoeuvring with difficulty". (*International Code).*
(ii) *Fog Signals*
B   The fog signal of a vessel being towed.
T   The signal of a power-driven vessel underway.
M   "My vessel is stopped and making no way through the water".
D   "Keep clear of me". The signal of a hampered vessel *e.g.* when towing another, fishing, or sailing.

SINGLE-LETTER SIGNALS WITH COMPLEMENTS

A with three numerals.   *Azimuth* or *Bearing* (True).

C with three numerals.   *Course* (True).

D with two, four or six numerals   *Date*. The first two figures indicate the day of the month. When they are used alone they refer to the current month. If two figures follow they indicate the month of the year. Where necessary the year may be indicated by two further figures).

G with four or five numerals.   *Longitude*. The last two numerals denote minutes and the rest degrees (measured from Greenwich).

K with one numeral.   I wish to *Communicate* with you by . . . (Complements Table I).
Example:
   K3 = by loud hailer (megaphone).

L with four numerals.   *Latitude*. The first two denote degrees and the rest minutes.

R with one or more numerals.   *Distance* (*R*ange) in nautical miles.

S with one or more numerals.   *Speed* in knots.

T with four numerals.   *Local Time*. The first two denote hours and the rest minutes.

V with one or more numerals.   *Speed* in kilometres per hour. (*V*elocity).

Z with four numerals.   Greenwich Mean Time. The first two denote hours and the rest minutes.

There is also a Single-letter code for signals between an ice-breaker and assisted vessels.

TWO LETTER-SIGNALS
Are not so commonly used as single-letter signals.

Examples:

UW   I wish you a pleasant voyage.

*UW*1   Thank you very much for your co-operation. I wish you a pleasant voyage.

RY   You should proceed at slow speed when passing me (or vessels making this signal).

*ZD*2   Please report me to Lloyds.

Complements — Complements express:

(a) Variations in the meaning of the basic signal.
    Example: DN = I have found the boat/raft.
          *DN*1 = Have you seen or heard anything of the boat/raft?

(b) Questions concerning the same basic subject or basic signal.
    Example: DY = Vessel (name or indentity signal) has sunk in lat . . . , long . . .
          *DY*4 = What is the depth of the water where vessel sank?

(c) Answers to a question or request made by the basic signal.
    Example: HX = Have you received any damage in collision?
          *HX*1 = I have received serious damage above the water line.

(d) Supplementary, specific or detailed information.
    Example: IN = I require a diver.
          *IN*1 = I require a diver to clear propeller.

THREE-LETTER SIGNALS
Beginning with M etc.

Examples:
MAC   I request you to arrange hospital admission.
MCX   Patient is delirious.

DISTRESS SIGNALS
"I am in distress and require assistance".

(Annexe IV from the *International Regulations for Preventing Collisions at Sea).*

(a) A gun or other explosive signal fired at intervals of about a minute.

(b) A continuous sounding of any fog-signal apparatus.

(c) Rockets or shells, throwing red stars fired one at a time at short intervals.

(d) A signal made by radiotelegraphy or by any other signalling method consisting of the group SOS in the Morse Code.

(e) A signal sent by radiotelephony consisting of the spoken word "Mayday" (corresponding to the French pronunciation of the expression *m'aidez, i.e.* "help me").

(f) The International Code Signal of distress indicated by *NC*.

(g) A signal consisting of a square flag having above or below it a ball or anything resembling a ball.

(h)   Flames on the vessel (as from a burning tar barrel, oil barrel, etc.).
(i)   A rocket parachute flare or a hand flare showing a red light.
(j)   A smoke signal giving off a volume of orange-coloured smoke.
(k)   Slowly and repeatedly raising and lowering the arms outstretched to each side.
(l)   The radiotelegraph alarm signal.
(m)   The radiotelephone alarm signal.
(p)   Signals transmitted by EPIRBS.
(o)   Approved signals transmitted by radiocommuncation systems.

RADIOTELEPHONY DISTRESS AND SAFETY SIGNALS

Mayday (Distress) procedure is used only if *immediate* assistance is required.

Use plain language when possible. The word **Interco** indicates that the message will be in the *International Code*.

(1)   If possible transmit the **Alarm Signal** (*i.e.* a two-tone signal for 30 to 60 seconds), but do not delay if there is insufficient time.
(2)   Ensure transmitter is switched to 2182 kc/s.
(3)   Then say:
　　　MAYDAY, MAYDAY, MAYDAY
　　　THIS IS .... (Ship's name or call sign 3 times) ...
　　　MAYDAY followed by ship's name or call sign
　　　POSITION ....
　　　NATURE OF DISTRESS ....
　　　AID REQUIRED .... **OVER.**
(4)   Listen for a reply and if none heard **repeat** above procedure, particularly during the 3-minute silence period commencing at each hour and half-hour.

Examples:

"Mayday Mayday Mayday. This is Nonsuch Nonsuch Nonsuch. Position fifty-four twenty-five north, sixteen thirty-three west. I am on fire and require immediate assistance. Over".

Pan (Urgency) — Any message prefixed with the word PAN indicates that an *urgent* message concerning the safety of a vessel, aircraft or person is about to be made.

Sécurité — (Safety) — Any message prefixed with the word SAY-CUR-EE-TAY indicates that a message concerning the safety of navigation or giving important meteorological warnings is about to be made.

The Radiotelephony Alarm Signal — Consists of a distinctive warbling note, alternately high and low in pitch and transmitted for a period of from 30 seconds to one minute. It can be generated by a special whistle and is easily distinguished by ear even through heavy interference. It is used to actuate special alarm receiving apparatus. It is used *only* to precede distress calls and should be followed at once by the Distress Call and Message. It should not be used if their is insufficient time. Transmission should be on 2182 Kc/s or for VHF to the International Safety Channel 16.

The Radio Telegraphy Alarm Signal — Consists of 12 dashes sent in one minute. The duration of each dash is 4 seconds, and the duration of the interval between two dashes 1 second. It is used to actuate special alarm receiving apparatus. It should be followed immediately by the Distress Call and Message. Transmission should be on 500 Kc/s.

The Distress Message should be followed by two dashes, each of about 10 seconds duration, to enable listening ships to take a radio bearing of the distressed vessel.

Responsibilities — If, after broadcasting a Distress Signal a vessel finds that she no longer needs assistance, it is important that she should inform all radio stations (both ships and shore) within range, otherwise vessels may be unnecessarily diverted from their course.

Every vessel has a statutory, as well as a moral, obligation to go to the assistance of a vessel in distress, if it is possible for her to do so.

LIFE-SAVING SIGNALS

(1) *Landing signals for the guidance of boats with persons in distress.* (Issued by Department of Transport).

This is the best place to land.

Vertical motion of arms, white flag, white light or flare. (Think of nodding the head affirmatively).

Or a *Green* star signal, or *K* given by light or sound. (*K*ome on!)

A steady white light or flare at a lower level and in line with the observer may be given to indicate direction of approach.

Landing here highly dangerous.

Horizontal motion of arms, white flags, light or flare. (Think of shaking the head negatively).

Or a *Red* star signal, or *S given by light or sound. (S*top!)

Landing here dangerous. Go in the direction indicated.

Horizontal motion as before followed by placing the white flag, light or flare on the ground and the carrying of another white flag, light or flare in the direction indicated. Or a red star fired vertically and a white star in the direction towards the better landing place. Or *S* followed by *R* if the better landing place lies to the right of the boat, or *L* if it lies to the left.

(2) *Signals to be employed in connection with the use of shore life-saving apparatus.*

Affirmative (in general): specifically; "Rocket line is held; Tail block is made fast; hawser is made fast; man in breeches buoy; Haul away".

Vertical motion of arms, white flag, light or flare, or a green star signal.

Negative (in general): Specifically; "Slack away; Avast hauling".
Horizontal motion of arms, white flag, light or flare, or a red star signal.
(3) *Replies from life-saving stations or rescue units to distress signals.*
You are seen — assistance will be given as soon as possible.
Orange smoke signal or combined light and sound signal consisting of 3 single white star rockets (which may sound) fired at intervals of approximately 1 minute.

SIGNALS MADE BY AIRCRAFT TO DIRECT SURFACE CRAFT TOWARDS AIRCRAFT OR SURFACE CRAFT IN DISTRESS

(1) The aircraft will *circle* the ship at least once.
(2) The aircraft will *cross ahead* of the ship at low altitude, opening and closing her throttle.
(3) The aircraft will *head* in the direction in which the ship is to be directed.

If assistance of a ship is no longer required, the aircraft will cross the ship's wake, close astern, at low altitude, opening and closing her throttle.

PILOT SIGNALS
"I want a pilot".
In the daytime:
The International Code signal *G* by flag or flashing or sound.
At night:
(1) The International Code signal *G* by flashing or sound.
(2) A bright white light, flashed or shown at short intervals just above the bulwarks for about a minute at a time.

A pilot vessel on her station flies flag *H* or a large white and red flag divided horizontally. At night she shows a white light over a red light in addition to side and stern lights, or anchor lights, and a bright intermittent light or flare.

QUARANTINE SIGNALS
Shown on arrival at a foreign port.
In the daytime:
Flag *Q* = *My vessel is "healthy" and I request free practique.*
Flag Q over 1st Substitute (*i.e. QQ*) I require health clearance.
At night:
Red light over white light, 2 metres apart, shown *only* within the port = I have not yet received practique.

LOCAL SIGNALS
Local signals are in force in many parts of the world, but should not be mistakable for International Signals.

Flag signals for controlling yacht races are used in many places. The following are common:

P = Preparative. Hoisted 5 minutes before the start
   of a race.
N = Negative. Races abandoned.
S = Shorter course to be followed.
1st Substitute = General recall.

A Sound Signal used in the Thames and Solent areas is another example:

4 short blasts: "I am about to turn — either completely round, or across the channel", followed by 1 short blast, if turning to starboard, or 2 short blasts, if turning to port.

FLAG SIGNALLING

As a general rule only one hoist of flag should be shown at a time. Each hoist or group of hoists should be kept flying until it has been answered by the receiving station. When more groups than one are shown on the same halyards they must be separated by a tackline (a length of halyard about a fathom long). The transmitting station should always hoist the signal where it can most easily be seen by the receiving station, that is, in such a position that the flags will blow out clear and be free from smoke.

Decimal Point — The Answering Pennant between numeral pennants indicates the decimal point.

How to Call — The identity signal of the station(s) addressed is to be hoisted with the signal. If no identity signal is hoisted it will be understood that the signal is addressed to all stations within visual signalling distance. If it is not possible to determine the identity signal of the station to which it is desired to signal, the group

"*VF* = You should hoist your identity signal" or

"*CS* = What is the name or identity signal of your vessel (or station)?"

should be hoisted first; at the same time the station will hoist its own identity signal. The group

"*YQ* = I wish to communicate by . . . (Complements Table 1) with vessel
       bearing . . . from me".

can also be used.

How to Answer Signals — All stations to which signals are addressed or which are indicated in signals are to hoist the Answering Pennant at the dip (about half the full extent of the halyards) as soon as they see each hoist and close up (to the full extent of the halyards) immediately they understand it; it is to be lowered to the dip as soon as the hoist is hauled down in the transmitting station, being hoisted close up again as soon as the next hoist is understood.

How to Complete a Signal — The transmitting station is to hoist the Answering Pennant singly after the last hoist of the signal to indicate that the signal is completed. The receiving station is to answer this in a similar manner to all other hoists.

How to Act when Signals are not Understood — If the receiving station cannot clearly distinguish the signal made to it, it is to keep the Answering Pennant at the dip. If it can distinguish the signal but cannot understand the purport of it, it can hoist the following signals:

"*ZQ = Your signal appears incorrectly coded. You should check and repeat the whole*", or

"*ZL = Your signal has been received but not understood*".

The Use of Substitutes — The use of substitutes is to enable the same flag — either alphabetical flag or numeral pennant — to be repeated one or more times in the same group, in case only one set of flags is carried on board. The First Substitute always repeats the uppermost signal flag of that class of flags which immediately precedes the Substitute. The Second Substitute always repeats the second and the Third Substitute always repeats the third signal flag, counting from the top of that class of flags which immediately precedes them. No Substitute can be used more than once in the same group. The Answering Pennant when used as a decimal point is to be disregarded in determining which Substitute to used.

Examples:

The signal "*VV* . . . Ice patrol is not on station" would be made as follows:

    V
    First Substitute

The number 1100 would be made by numeral pennants as follows:

    1
    First Substitute
    0
    Third Substitute

The signal "*L2330 =* Latitude 23° 30" would be made as follows:

    L
    2
    3
    Second Substitute
    0

In this case the Second Substitute follows a numeral pennant and therefore it can only repeat the second numeral in the group.

How to Spell — Names in the text of a signal are to be spelt out by means of the alphabetical flags. The "*YZ = The words which follow are in plain language*" can be used if necessary.

THE MORSE CODE

The Morse code consists of groups of dots and dashes, or shorts and longs, each group representing a letter or number, which can be made by flashing a light, by sound, or by using a flag. The dots should be made as short as possible (so long as they can be clearly seen or heard), and the dashes should be at least three times as long as the dots. A pause is made between each symbol.

The Morse Alphabet — In the Morse alphabet the letters which occur most frequently in English have been given the shortest symbols.

A	for Alpha	‑ —	N	for November	— ‑
B	Bravo	— ‑ ‑ ‑	O	Oscar	— — —
C	Charlie	— ‑ — ‑	P	Papa	‑ — — ‑
D	Delta	— ‑ ‑	Q	Quebec (Kibbeck)	— — ‑ —
E	Echo	‑	R	Romeo	‑ — ‑
F	Foxtrot	‑ ‑ — ‑	S	Sierra	‑ ‑ ‑
G	Gulf	— — ‑	T	Tango	—
H	Hotel	‑ ‑ ‑ ‑	U	Uniform	‑ ‑ —
I	India	‑ ‑	V	Victor	‑ ‑ ‑ —
J	Juliet	‑ — — —	W	Whisky	‑ — —
K	Kilo	— ‑ —	X	X-ray	— ‑ ‑ —
L	Lima	‑ — ‑ ‑	Y	Yankee	— ‑ — —
M	Mike	— —	Z	Zulu	— — ‑ ‑

The words placed against each letter are used to prevent mistakes arising from the similar sound of letters. For example: the man saying "n,d" may be mistaken as saying "m,b", but "november, delta", is unmistakable.

Numerals — The numeral always consists of five elements — dots or dashes, so that the receiver is able to check the number by counting both the dots and dashes.

1	‑ — — — —	Unaone	pronounced	Ooh-nah-wun
2	‑ ‑ — — —	Bissotwo	pronounced	Bees-soh-too
3	‑ ‑ ‑ — —	Terrathree	pronounced	Tay-rah-tree
4	‑ ‑ ‑ ‑ —	Kartefour	pronounced	Kar-tay-fower
5	‑ ‑ ‑ ‑ ‑	Pantafive	pronounced	Pan-tah-five
6	— ‑ ‑ ‑ ‑	Soxisix	pronounced	Sok-see-six
7	— — ‑ ‑ ‑	Setteseven	pronounced	Say-tay-seven
8	— — — ‑ ‑	Oktoeight	pronounced	Ok-toh-ait
9	— — — — ‑	Novenine	pronounced	Noh-vay-niner
0	— — — — —	Nadazero	pronounced	Nah-dah-zay-roh

The Department of Transport Examinations require a speed of 6 words per minute for Morse. To test your speed in sending and receiving, note the number of seconds you take to send or receive 20 words. Divide the number of seconds into 1200 and the result is the speed in words per minute.

Procedure Signals — A bar over the letters composing a signal denotes that the letters are to be made as one symbol.

$\overline{AA}\ \overline{AA}\ \overline{AA}$, etc.	*Call for unknown station or general call to all stations within visual signalling distance (for flashing light)*. The call is continued until the station addressed answers. Answered by $\overline{TTTTT}$, etc.
CQ	*Call for unknown station(s) or general call to all stations* (for flags, radiotelephony and radiotelegraphy). When spoken "Charlie Quebec".
K	*"I wish to communicate with you" or "Invitation to transmit".*
CS	*"What is the name or identity signal of your vessel (or station)?"*
DE	*"From . . ."* (used to precede the name or identity signal of the calling station).
AS	*Waiting Signal.* When made at the end of a signal indicates that the other station must wait for further communications.
	Period Signal. When inserted between groups it serves to separate them to avoid confusion.
C	*Affirmative* — YES or "The significance of, the previous group should be read in the affirmative".
N	*Negative* — NO or "The significance of the previous group should be read in the negative. When used in voice transmission the pronunciation should be "NO".
RQ	*Interrogative*, or "The significance of the previous group should be read as a question".
	C, N and *RQ* cannot be used in conjunction with single-letter signals.
	When *N* or *No* and *RQ* are used to change an affirmative signal into a negative statement or into a question, respectively, they should be transmitted *after* the main signal. (*R*ead as *Q*uestion).
	Example: DN N = I have not found the boat/raft.
	JA RQ = "Do you require fire-fighting appliances?"

OK	Acknowledging a correct repetition or "*It is correct*".
$\overline{AAA}$	*Full stop* or *decimal point*.
$\overline{EEEEE}$, etc.	*Erase Signal* used to indicate that the last group or word was signalled incorrectly. It is to be answered with the Erase signal. When answered, the transmitting station will repeat the last word or group which was correctly signalled and then proceed with the remainder of the transmission.
RPT	*Repeat Signal.* Made by the transmitting station to indicate that it is going to repeat ("I repeat"). If such a repetition does not follow immediately after *RPT* the signal should be interpreted as a request to the receiving station to repeat the signal received ("Repeat what you have received"). Made by the receiving station to request for a repetition of the signal transmitted ("Repeat what you have sent").
AA	"*All after* . . ." (used after *RPT*) means "Repeat all after . . .".
AB	"*All before* . . ." (used after *RPT*) means "Repeat all before . . .".
WA	"*Word or group after* . . ." (used after *RPT*) means "Repeat word or group after . . .".
WB	"*Word or group before* . . ." (used after *RPT*) means "Repeat word or group before . . .".
BN	"*All between* . . . and . . ." (used after *RPT*) means "Repeat all between . . . and . . .".
T	"*Word or group received*" (used in flashing light).
R	"*Received*" or "I have received your last signal".
$\overline{AR}$	*Ending signal* or End of Transmission signal. Answered by *R*.

FLASHING LIGHT SIGNALLING

Signalling by flashing at a distance of several miles can be carried out by night *and* day, if a daylight Signalling Lamp (an **Aldis lamp**) is used.

A signal made by flashing light is divided into the following parts:

The Call — Consists of the General call ($\overline{AA}\ \overline{AA}\ \overline{AA}$), etc., or the identity signal of the station to be called. Answered by the Answering Signal $\overline{TTTTTTT}$, etc.

The Identity — The transmitting station makes *DE* followed by its identity signal or name. This is repeated back by the receiving station which then signals its own identity signal or name. This is also repeated back by the transmitting station.

The Text — Consists of plain language or code groups. When code groups are used they should be preceded by the signal *YU*. Receipt of each word or group is acknowledged by *T*.

The Ending — Consists of the Ending Signal $\overline{AR}$ which is answered by *R*.

Note: The Call and Identity may be omitted when two stations have established communications and have already exchanged signals.

Ships passing at night often exchange their names and the ports whence and to which they are bound. The procedure is simple. For example, the S.S. *Strathmore* meeting an unknown ship makes the call. When answered she makes "Strathmore London to Gibraltar *AR*". The other ship, having answered each word with *"T"* and the Ending Sign with *"R"* makes the Call Sign following it with "Chantala Genoa to Dunkirk" probably adding "Bon Voyage" or "Good night".

SOUND SIGNALLING

Sound signalling is necessarily slow since it is made by whistle, siren, fog-horn, etc. The misuse of sound signalling may create serious confusion at sea and sound signalling in fog should therefore be avoided. Signals other than those required by Rules 34 and 35 should be used only in extreme emergency and never in busy navigational waters.

RADIOTELEPHONY

Calling — The call sign or name of the station called or *CQ (Charlie Quebec)* for all stations in the vicinity, not more than three times.

DE (Delta Echo) — The call sign or name of the calling station not more than three times.

Replying — The reply to calls consists of the call sign or name of the calling station, not more than three times.

DE (Delta Echo) — The call sign or name of the station called, not more than three times.

Interco indicates that Code groups of the *International Code of Signals* are following. Words of plain language may also be in the text when the signal includes names, places, etc. In this case the group *YZ (Yankee Zulu)* is to be inserted if necessary.

Procedure Signals are to be spelt in accordance with the spelling tables, *e.g.:*

    AS (Alpha Sierra); AR (Alpha Romeo); R (Romeo).

For radiotelephony or loud hailer:

IN-TER-CO	International Code group(s) follow(s).
STOP	Full Stop.
DAY-SEE-MAL	Decimal Point.
KOR-REK-SHUN	Cancel my last word or group. The correct word or group follows.

SEMAPHORE

Semaphore is no longer practiced or examined in the Merchant Navy.

The letters of the semaphore alphabet are shown in the figure. Great care must be taken to make the angles clearly and correctly. Arm, wrist and flag should form a straight line, and, to aid this to be carried out, the forefinger should lie along the stave of the flag. When sending, get into a conspicuous position and choose a background which is as far away as possible, the sky being the best. Semaphore is much quicker than Morse signalling by hand-flags or arms.

"I wish to communicate by semaphore", K1 or, if close enough to be seen, the Attention sign.

On receipt of the call, the station addressed should hoist the Answering Pennant at the dip, or make the Answering sign, or, if unable to communicate by semaphore, should reply with the signal *YS1*.

The sender will make the attention sign and wait until the Answering Pennant is hoisted close up, or the Answering sign is made, commencing transmission after a reasonable pause.

The signal should always be made in plain language and numbers spelt out in words.

At the end of each word the arms are to be dropped to the break position. When double letters occur, the arms are to be dropped to the break position after the first letter is made and then moved out to the second letter without pausing.

The Erase signal is a succession of *E*s.

Each word in answered by the receiving station making the letter *C*.

If this letter is not made the word is to be repeated.

All signals end with the Ending signal *AR*.

Single Up — To take in all doubled mooring lines before leaving a quayside.

Skates — These fenders are fitted to the inboard side of a lifeboat's hull to assist its progress down the ship's side if she has a list.

Skiff — A light transom-sterned boat, usually 16 or 18 feet long.

Slew — To turn anything round.

Sling — A rope with its ends short-spliced together used to lift cargo.

Snotter — A length of rope or wire with an eye spliced in each end; used for slinging bales, etc.

Soldier's Wind — A fair wind to and from a place, so that there is no need to tack.

Sound — To find the depth of water. Bilges and tanks are sounded morning and evening by the carpenter.

A	B	C ANSWERING SIGN	D	E
F	G	H	I	J
K	L	M	N	O
P	Q	R	S	T
U	V	W	X	Y
	Z	ATTENTION	BREAK	

Spanish Windlass — Used to heave two ropes together. Take a round turn round both ropes with a line; lay a bar across the rope; hitch the ends of the line to two marline-spikes and, using them as levers, twist them opposite ways round the bar, thus tautening the line and bringing the two ropes together.

Spring Tides — Tides which rise and fall more than normally. They occur every fortnight at the new and full moon.

Sprung — Cracked or splintered.

Spritsail — A sail used as a mainsail in Thames barges and sometimes in ships' boats. Instead of having a gaff it is supported by a sprit, one end of which is fitted into the peak of the sail and the other supported by a stout snotter near the foot of the mast. Spritsails are furled by brails.

Spunyarn — Ropeyarns laid up together and tarred with Stockholm tar. Used for servings, light seizings, etc.

S.S. — Steam ship.

Fig. 17 — Slinging a Stage.

Stage — A stage consists of a long plank fitted with a short cross piece at each end and is used when painting the side. **To Sling a Stage** — If a single end of rope is to be used, first make a bowline at one end with a loop about 6 feet long. Form a marline-spike hitch at the end of the loop. Slip the marline-spike hitch over the end of and cross piece of the stage so that the middle part is on top of the stage and the two outer parts below on each side of the cross piece, then pull taut.

Stand-By — A caution to make ready.

Starboard — The right-hand side of the ship when looking forward. Fittings on the starboard side are always numbered with *odd numbers.*

Stay — To go about. **Miss Stays** — To fail to get from one tack to the other. **Slow in Stays** — Slow in going about.

Steering — When the top of the steering wheel is moved to starboard, the rudder is turned to starboard and, if the ship is moving ahead through the water, the ship's head swings to starboard. When the top of the wheel is moved to port, the rudder moves to port and the ship's head moves to port, and her stern to starboard.

In the compass bowl is a black line called the **lubber line** which is exactly in line with the ship's head. When the lubber line swings out to port of the course you are steering, turn the wheel to starboard to bring the lubber line back to the course. A ship does not answer her rudder immediately, so when the wheel is put over it must be left over a little time before the ship will start to swing back. Directly she begins to swing the wheel should be eased. The art of steering is to use as little helm as possible and to turn the wheel directly the ship's head begins to swing; it requires great care and practice. By watching the land or clouds ahead of the ship you can often tell when the ship's head begins to swing before the compass shows it.

When relieved at the wheel tell the helmsman the course to be steered (he should repeat it to show that he has heard correctly) and the amount of helm the ship is carrying to keep her on her course, and report the course you have been steering to the officer of the watch.

STEERING ORDERS

"Starboard" — The top of the wheel, rudder and the ship's head all move to starboard. In a boat the starboard yoke line is pulled and rudder and boat's head move to starboard, but if steering with a tiller, the tiller is put to *port*.

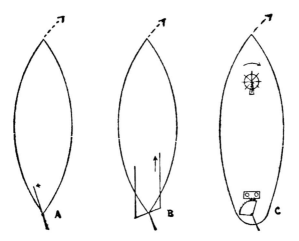

Steering by (*A*) tiller. (*B*) Yoke and yoke lines. (*C*) Wheel and steering engine. The figure illustrates the order "starboard". In each case the rudder and the vessel's head move to starboard. The top of the wheel moves to starboard, the starboard yoke line is pulled, but the tiller is pushed over to *port*. If the vessel had sternway on her stern would move to starboard.

"Port" — Wheel, rudder and ship's head move to port, port yoke line is pulled and tiller is put to *starboard*.

The amount that the rudder is to be moved is often given as well (in multiples of 5°), and the wheel is then turned until the wheel indicator shows that the right number of degrees has been given, *e.g.,* **port twenty-five** means that the rudder should be turned until it makes an angle of 25° with the fore and aft line of the ship, and since the wheel is usually geared so that one complete revolution turns the rudder through 10° the indicator will show 25° when the wheel has been turned through $2\frac{1}{2}$ revolutions. At other times the order **hard-a-port**, or **hard-a-starboard**, may be given, the wheel being turned until the rudder is as far over as it will go (about 35°), but it should never be jammed over. About two complete turns of a yacht's wheel will put the rudder hard over.

"Ease the Helm" or **"Ease the Wheel"** — The amount of helm on is then reduced or eased.

"Midships" — The wheel is put amidships.

"Meet Her" — Is ordered when the ship is swinging too rapidly. The wheel is then put in the opposite direction to check, or meet, her.

"Steady" or **"Steady as you Go"** — The ship is to continue in the direction in which she is heading at the time the order is given.

"Nothing to Starboard" (or Port) — Is sometimes given owing to some danger on that side, and the helmsman must not let the ship swing to starboard (or port) of the course he is steering.

All orders should be repeated exactly as given, and when the purport has been carried out it should be reported.

Examples:

Order	Reply	Purport carried out
Starboard ten	Starboard ten	Ten of starboard on
Hard-a-port	Hard-a-port	Hard-a-port
Midships	Midships	Wheel's amidships
Steady	Steady	Steady three three zero
		(*i.e.* the compass heading)

All orders remain in force until a fresh order is given.
For steering under sail, *see* "Sails and Sailing".
For the effect of the propeller upon steering *see* "Handling Craft".

Stiff — A vessel not easily heeled over is said to be stiff.
Still — The order "Still" is given when it is necessary to stop some order from being carried out. Each man stops what he is doing, stays still

and keeps silent. If the order is given when a boat is being lowered the lowerers immediately take another turn with the falls or belay them. Operations are resumed at the order "Carry on".

Stopper — A length of rope used for temporarily making a rope fast. Stoppers for wire ropes are made of chain with a rope tail. To pass a stopper, take a half-hitch round the rope against the lay and then dog the ends round and round the rope with the lay.

Stove-In — Broken in; holed.

Strake — A plank in a boat's side or a length of plating in a ship's hull.

Stream the Log — To pay the rotator and logline overboard.

Striker — A paint brush on a long handle.

Strop — A short length of rope with its ends spliced together.

Strum Box — A grating built round the end of a bilge suction pipe to keep the rose on the pipe's end from becoming choked.

Sujee — Powder used for cleaning paintwork.

Superstructure — Parts of the ship built above the upper deck.

Surge — To render or ease a hawser round a winch barrel or capstan.

Swell — Waves caused by a recent wind or a wind at a distance.

Swifter — One leg of a pair of shrouds; a line binding in the bars of a capstan.

Swig — To take an extra pull on a halyard, after it has been turned up on the cleat, by pulling at right angles to it.

T

"T is for topsails, topgallants also".

Tabernacle (1) — A frame between mast-thwart and step which takes the foot of the mast. (2) A hinge which allows the mast to be lowered.

Tackline — A length of rope about 6 feet long, used to separate flags which, if hoisted at the usual distance apart, would convey a different meaning to that intended.

Take To — To take turns with a hawser round a winch or capstan barrel ready for heaving in.

Taut — Tight.

Telephone — When receiving orders by telephone or voice pipe, always repeat them back again so that the man passing the orders knows that you have heard it correctly.

Tender — A vessel easily heeled over is said to be tender. (2) A boat or small vessel used to attend on her parent ship.

T.E.U. — Twenty foot Equivalent Unit. A container 20 feet long, 8 feet wide and 8 feet high. The size of a containership may be given as a number of TEU's.

Thermometer — An instrument for measuring temperature. In the Celsius (Centigrade) scale water freezes at 0°C and boils at 100°C. In the Fahrenheit scale water freezes at 32°F and boils at 212°F. Average summer temperature is 17°C and 62°F.

Thimble — A metal ring set in the eye of a rope or a cringle of a sail or awning.

Thole Pins — Pegs shipped in the gunwale, instead of crutches, when rowing.

Toggle — A short wooden peg fitted at the head of a flag, round which the halyard is made fast. Any small peg.

Tonnage — Gross and net register tonnage represent space, not weight. Gross tonnage = the number of tons enclosed in a ship (1 ton = 100 cubic feet). Net tonnage — the cubic capacity, in tons, of the *earning* space in a ship (*i.e.* holds and passenger accommodation. Deadweight tonnage = the number of tons weight a ship can carry. Displacement tonnage = the number of tons of water which a ship displaces, *i.e.,* her total weight. Gross tonnage is about 60% of the Standard Displacement.

Trimaran — A vessel with an outrigger and float on each side.

Trick — A turn, spell or period of duty. A trick at the wheel or look-out lasts for two hours.

Trim — The manner in which a vessel sits on the water.

Trinity House — The Corporation which controls the lighthouses and lights around the English coasts.

Truck — The circular cap at the top of a mast.

Tumble-home — The leaning inwards of a vessel's sides above the water.

Tumbler — A device for letting go from inboard the outboard chock on which a lifeboat rests.

Twine — Supplied in skeins. Seaming twine for sewing canvas and roping twine (which is thicker) for whippings.

Two Blocks — A tackle is said to be two blocks when the blocks have been hauled close together.

U

"U is the Union which flies at the fore".

(Probably the white-bordered Union Jack which, when hoisted at the foremast head, was a signal for a pilot).

Unship — To take off, to remove.

V

"V is the Vane which detects every flaw".
(The wind-vane at the truck watched by the helmsman).

Vast — Avast, stop.
Veer — To pay out or slack away cable. When the wind shifts from left to right, or clockwise, it is said to veer.

W

"W, the Wheel where we all take a turn".

Waist — Amidships.

Wake — The track left by a ship.

Walk Back — To pay out a hawser or cable by reversing the winch or capstan.

Warp — To haul a vessel into position with a hawser.

Warps — Ropes which hold a vessel when moored.

Watches — Are of four hours, changing at 8, 12 and 4 o'clock. (*See* also "Bells"). The evening watch is divided into two: the first and last dog watches. This is done in order that the hands shall not get the same watches every day. Seamen are usually divided into three watches named Red, White and Blue, or A, B and C. Men not included in the watches are known as "Daymen". Seamen are occasionally divided into two watches, named Port and Starboard. In Her Majesty's ships the seamen of each watch are also divided into parts of the ship, *viz.* Forecastlemen (FX), Foretopmen (FT), Maintopmen (MT) and Quarterdeck (QX), usually forming the guns' crews and working in their own parts of the ship. Each part of the ship is under a petty officer known as the "Captain of the Top".

Waterlogged — Full of water, but floating.

Watertight Doors — Are placed in bulkheads (except in the collision bulkhead which never has any openings). Besides being watertight they are very strong, to withstand the pressure of water, and are shut by means of a wheel and gearing worked from the deck above. In large ships the lower and more important watertight doors can be shut from the bridge by remote control.

Way — A vessel is under way when she is not at anchor, made fast to the shore or aground. When she begins to move through the water she *gathers way*, if moving ahead she has *headway*, if astern, *sternway*. *Steerage way* to move through the water sufficiently fast for the rudder to have an effect.

Weather — The description of the weather in log books and weather reports is usually shortened by the use of the following letters and numbers. Wind is named by the point of the compass *from* which it blows, and by a Beaufort Number named after Admiral Beaufort (who invented them).

Beaufort Number	Average Velocity in Knots	Description
0	0	Calm
1	2	Light airs
2	5	Light breeze
3	9	Gentle breeze
4	13	Moderate breeze
5	18	Fresh breeze
6	24	Strong breeze
7	30	Near Gale
8	37	Gale
9	44	Strong Gale
10	52	Storm
11	60	Violent Storm
12	over 68	Hurricane

Letters to Indicate the State of the Weather

b   Blue sky (less than a quarter covered)

bc Sky partly cloudy (between one quarter and three quarters covered)

c   Generally Cloudy (more than three quarters covered)

d   Drizzle or fine rain

e   Wet air without rain

f   Fog

g   Gale

h   Hail

kg Line squall

kz Sand or dust storm Lightning

m   Mist

o   Overcast sky (whole sky covered with one impervious cloud)

p   Passing showers

q   Squalls

r   rain

rs Sleet, *i.e.,* rain and snow together

s   Snow

t   Thunder

tl Thunderstorm

u   Ugly, threatening sky

v   Exceptional visibility

w   dew

z   Dust haze

Capital letters are used to indicate "intense".

Weather Rhymes:

> Red sky in morning
> Sailor's warning
> Red sky at night
> Sailors' delight

> When the wind's before the rain
> Soon you can make sail again,
> When the rain's before the wind
> Your topsail halyards you must mind.

> *A shift of wind*
> When the wind shifts against the sun
> Trust it not, for back it will run.

With a low barometer

First rise after a very low	Mackerel sky and mare's tails
Indicates a stronger blow.	Make lofty ships carry low sails.

> Long foretold — long past,
> Short notice — soon past.

> With rising wind and falling glass (barometer)
> Soundly sleeps a careless ass.

Weather Tide — A tide setting against the wind.

Weather Side — The side on which the wind is blowing. Never attempt to throw anything overboard on the weather side.

Weigh — To heave up the anchor.

Wing and Wing — A boat sailing with the wind aft, with one sail boomed out to port and the other to starboard is said to be running wing and wing, or goose-winged.

Wreck Marking — Wrecks are now marked by I.A.L.A. buoys which may be doubled for emphasis.

Wriggle or Eyebrow — A steel strip bent over a porthole or scuttle to prevent water trickling in.

W/T — Wireless telegraphy; radio.

"X, Y and Z is the name on our stern".

Y

Yarns — Twisted up threads of rope made of hemp, manilla, sisal, coir, etc. The yarns are laid up the opposite way into strands, and the strands laid up the opposite way again into rope.

Yaw — A vessel is said to yaw when her head swings first in one direction and then in another due to bad steering or a high quarterly sea.

Z

Zone Time — A method adopted by most countries for keeping their time a whole number of hours fast or slow of Greenwich time. The world is divided into 24 zones measuring 15° of longitude, Greenwich being in the middle of zone 0. The zones to the eastward of zone 0 are numbered −1 to −11, and the eleven to the westward from +1 to +12. For example, the time kept in zone −2 is two hours ahead of Greenwich time, that in zone +6, six hours behind Greenwich time. The zone exactly opposite zone 0 (on the other side of the world) is divided into two half-zones called −12 and +12.

USEFUL DATA

Length
1 Kilometre (Km) = 0·54 Sea Miles (M).
1 Mile (M) = 1·853 Km.
1 Metre = 0·547 fathoms.
1 Fathom = 1·829 metres.
1 Metre = 3·2809 feet.
1 Foot = 0·3048 metres.

Weight
1 Kilogram (Kg) = 2·2046 pounds (lbs).
1 Pound = 0·4536 Kg.
1 Tonne = 1000 Kg.
$\qquad$ = 0·9842 tons.
1 Ton = 1016·05 Kg.

Capacity
(1 Litre = 1000 cub. cm.).
1 Litre = 0·22 gallons.
1 Gallon = 4·546 litres.

Fresh Water
1 Gallon weighs 10 lbs.
1 Ton occupies 35·8 cubic feet.

Tanks
Volume of rectangular tank = l × b × d.
Volume of cylindrical tank = $l\pi r^2$
l = length, r = radius, $\pi = \frac{22}{7}$
Capacity of tank = vol. ÷ 35 (for salt water) gives capacity in tons.
Capacity of tank = vol. ÷ 36 (for fresh water) gives capacity in tons.
Volume × $6\frac{1}{4}$ gives number of gallons of fresh water.
Volume in cubic metres gives tonnes of fresh water.

Volume (approximate) of ship-shaped body
l × b × d × block coefficient. The block coefficient of a boat may be from 0·6 (for fine lines) to 0·8 (for full lines).

Area of Sails
The area of a triangle = $\frac{1}{2}$ height × base. Take the longest side as base. If the sail is 4-sided, measure a diagonal and find the area of the 2 triangles.

Rope
Breaking stress and Safe Working Loads. *See* **Rope and Rigging.**

Chain
Breaking stress of chain cable (in tons) = diameter of link (in inches) × 30.
Weight of chain in tons per fathom
$$\frac{d^2 \text{ inches}}{40}$$

Cement
For ordinary purposes mix 1 part cement to 3 parts sand.

Thermometer Scales
To convert Celsius to Fahrenheit, multiply by $\frac{9}{5}$ and add 32.
To convert Fahrenheit to Celsius, subtract 32 and multiply by $\frac{5}{9}$.

MY SHIP

Name ....

Port of Registry ....

Built ....

Signal Letters ....

Length ....

Breadth ....

Draft ....

Gross Tonnage ....

Nett Tonnage ....

Engines ....

No. of Crew ....